Health and Safety in Welding and Allied Processes

Health and Safety in Welding and Allied Processes

FIFTH EDITION

Jane Blunt and Nigel C Balchin

CRC Press
Boca Raton Boston New York Washington, DC

WOODHEAD PUBLISHING LIMITED
Cambridge England

Published by Woodhead Publishing Limited, Abington Hall, Abington
Cambridge CB1 6AH, England
www.woodhead-publishing.com

Published in North America by CRC Press LLC, 2000 Corporate Blvd, NW
Boca Raton FL 33431, USA

First published 1956, Institute of Welding
Revised and enlarged, July 1963
Second edition, 1965
Third edition, 1983, The Welding Institute
Fourth edition, 1991, Abington Publishing
Fifth edition, 2002, Woodhead Publishing Limited and CRC Press LLC

British Library Cataloguing in Publication Data
A catalogue record for this book is available from the British Library.

Library of Congress Cataloging in Publication Data
A catalog record for this book is available from the Library of Congress.

Woodhead Publishing ISBN 1 85573 538 5 ✓ 𝒬𝓗. 15.1.03
CRC Press ISBN 0-8493-1536-0
CRC Press order number: WP1536

Cover design by The ColourStudio
Typeset by SNP Best-Set Typesetter Ltd, Hong Kong
Printed by TJ International Ltd, Cornwall, England

Contents

Introduction

This is the fifth edition of this work. It has been extensively revised to take into account changes in technology and legislation. Every effort has been made to include the legislative requirements of both the United Kingdom and the United States of America in order to make this book useful to personnel on both sides of the Atlantic. References for each country are given throughout.

Some reorganisation of the contents has taken place, and a worked example has been included in Appendix B to illustrate the method of risk assessment, which is the basis for the assessment and control of risk in the United Kingdom.

The work begins with a description of the core safety requirements. It then describes the special hazards found in the welding environment – noise, radiation, fume, gases, etc, in terms of their effects and the strategies that might be adopted to avoid them. The central part of the book takes each major joining technology in turn, and discusses the key hazards that are most relevant to that technology. Finally there is a chapter on testing and welding in situations of increased hazard.

The information in this book is believed to be correct at the time of going to press. However, it must be stressed that the onus is on employers to address the risks that exist in their own workplaces, and to ensure that they are complying with the laws that govern work in their own locality.

This book should be of use to welders, their managers, and to all health and safety practitioners who have welding and similar processes taking place in their workplace.

Part 1

Risks and Principles for
their Control

1

Setting up the Workplace

In both the United Kingdom and the United States of America, there is a legislative framework that assigns a very large measure of responsibility to employers for the health and safety of their employees. The detailed approach is slightly different and readers need to familiarise themselves with the requirements. Where they have doubts, they should consult the enforcing authorities for advice:

- The Health and Safety Executive (United Kingdom)
- Occupational Safety and Health Administration (United States of America).

The general requirements in the United Kingdom are laid down in the Health and Safety at Work, etc, Act, 1974,[1] which places a duty on all employers to ensure as far as is reasonably practicable, the health, safety and welfare of all of their employees while they are at work. Many duties are also extended to those not in their employment but who may be affected by the employer's undertaking. The Act enabled the making of Regulations, which contain detailed specific requirements, which employers are required to comply with. The basis upon which employers should act is one of risk assessment – where employers must analyse the risks associated with their work activities and implement measures to control those risks.[2] Employees are required to cooperate with their employer's efforts to meet the requirements of the Act and the Regulations.

There are two useful websites where further information may be obtained, Her Majesty's Stationery Office,[3] where the full text of all Statutory Instruments published since 1987 is available to view and print, and the Health and Safety Executive,[4] (HSE), where there is a great deal of advice and guidance.

3

The general requirements in the United States of America are laid down in section 5 of the Occupational Safety and Health Act of 1970,[5] which requires employers to furnish each of their employees with employment and a place of employment which are free from recognised hazards that cause or are likely to cause death or serious physical harm to those employees. The Act requires employers to comply with the occupational safety and health standards that it promulgates. The Occupational Safety and Health Administration (OSHA) maintains a website from which access can be gained to Federal Regulations.[6] Employees are required to comply with the rules, regulations and orders that apply to them.

The net effect in both countries is that in order to ensure that the workforce remains safe and that the requirements are met, a system is needed to manage safety in the workplace. An efficient system will not only meet the legislative requirements, but is also cost effective in minimising lost time through illness and injury.

The employer should set up a policy for the assurance of health and safety and assign responsibilities for undertaking the many tasks that will need to be carried out. The workplaces will need to be constructed and maintained in good order. The work equipment will need to be fit for its purpose and properly maintained .[7] Setting up a safety committee enables worker participation and establishes good communication. Safety rules will be needed and the workforce will need to be trained so that they know what hazards they face, the preventive and protective measures that are needed to avoid the risk of injury or ill health, and how to make the best use of those measures, including personal protective equipment if it is needed.[8] An inspection programme will be needed to ensure that the measures are adequate and that tasks are being carried out as required. In many workplaces, there will be a need for some health surveillance and monitoring of key indicators.

The Workplace

First, the prescribed poster should be put up in the workplace. In the UK, this is available from the HSE or good bookshops.[9] Alternatively, the prescribed leaflet[10] may be distributed to every employee. In the USA, the prescribed poster[11] can be downloaded from the government website. The workplace should be in accordance with the provisions of the Workplace Regulations,[12] or in the USA, according to the requirements of subparts D and J of 29 CFR 1910.[13,14]

Indoor workplaces should be kept at a reasonable temperature. A temperature of 16 °C or above is recommended where personnel are undertaking light work, and a minimum of 13 °C where heavy work is undertaken. Measures may need to be taken in hot weather to prevent people from becoming overheated.

Adequate sanitary facilities should be provided, with facilities for washing and drying the hands. The facilities should be kept clean. An area should be set aside, separate from the work area, where food and drink can be consumed without contamination by substances hazardous to health.

Walkways should be marked and kept clear. The walkways should have surfaces that are free from holes, slippery substances and water, to avoid slips, trips and falls. There should be railings or other guards to prevent people from falling down stairs, shafts, etc.

Lighting

When work must be carried out in areas where insufficient daylight is available it will be necessary to provide artificial lighting, which will almost invariably be electric. Two cases must be covered: normal operation and emergency lighting. General advice is given in an HSE publication.[15]

Normal lighting

The information in Table 1.1 below has been selected from Table 1 of the now obsolescent British Standard[16] as that most likely to be applicable to welding activities.

The general run of welding work on mild steel plate, often with a black surface, will be of very low contrast. Although arc welding is an almost unique operation, in that the arc emits far more light than

Table 1.1. Illuminances and corresponding activities

Standard service illuminance (lux)	Visual task	Details to be seen	
		Size	Contrast
500	Moderately difficult	Moderate	Low
750	Difficult	Small	Low
1000	Very difficult	Very small	Very low
1500	Extremely difficult	Extremely small	Very low

any practicable artificial illumination, good general illumination will permit the use of a lighter shade of viewing filters because the eyes adapt to the general level of illumination by narrowing of the pupils and the arc light has to be reduced less to match. This gives welders a better view of the weld with less eye strain and renders them less susceptible to dazzle by an accidental view of an arc.

Good lighting is also important to facilitate preparatory work, such as edge preparation and assembly of components, and visual checks after welding by the welder himself, etc. Where there is rotating machinery (such as turntables for spraying, or lathes) the designer of the lighting system should avoid stroboscopic effects.

The environment in a normal welding shop will require allowance for reduction of output due to dust accumulation on luminaires (lighting fittings) during the intervals between routine lamp replacement and cleaning. It is not necessary to paint a welding shop black to avoid reflection of ultraviolet (UV) light (see Chapter 7).

For work on site, some welding generators are available with an outlet to power lights; as this is often of low power or of non-standard voltage or frequency, the exact facilities required should be checked against the specification.

For the illumination of fuel gas stores, where a leak could give rise to an explosive atmosphere, flameproof equipment will be required (see Chapter 4), unless it is possible to site the lighting outside the hazard area. This may offer security advantages.

Emergency lighting

If a complete electrical power supply failure occurs after dark, emergency lighting will be needed to ensure that workers are able to see well enough to carry out such actions as the following:

1 Making safe any radiographic equipment, especially isotope sources,
2 Shutting down all gas flames for welding cutting preheating, etc,
3 Switching off all electric welding equipment,
4 Rendering safe any equipment relying on supplies also cut off by an electric power failure, such as water cooling, compressed air or ventilation systems,
5 Ensuring that all crane motors are switched off and that any suspended loads which present a hazard, will be marked if necessary,

6   Rescuing anyone trapped, such as in a crane jib or lift,
7   Evacuating the premises in an orderly fashion, making sure that
    no one is left behind.

If it is necessary to cut off the supply in the event of fire, similar
considerations will apply. Escape lighting should:

1   Indicate the escape routes clearly and unambiguously
2   Illuminate those routes that allow safe exit
3   Enable the ready location of fire alarm call points and fire fight-
    ing equipment on escape routes.

On defined escape routes, 0.2 lux illumination is required and 1 lux
where they are not defined, that is, where they run across an
open area. Regular servicing, inspection and testing must be organised to make sure that the system will function if and when it is
required.

Housekeeping

The workplace should be kept clean and tidy. Trip hazards can be
avoided by careful siting of leads and hoses and not putting tools
down where people may walk. Tools should be put away each day.
Oily waste should be placed in metal bins. All bins should be
emptied regularly to avoid an accumulation of combustible waste.
Where personal protective equipment is provided, there should be
provision for its safe storage, and it should be put away when not
in use. Accumulations of metal dust, which are especially likely in
a thermal spray workshop, can be explosive.

Manual Handling

Many injuries are attributable to manual handling.[17] Lifting tasks
should be assessed critically. Manual handling tasks that are likely
to be hazardous should be avoided where possible. Many can be
avoided by the use of suitable lifting aids, such as trolleys and sack
barrows. Where manual handling is essential, the task should be
assessed and personnel should be trained in good lifting technique.
Good practices include bending the knees rather than the back to
pick up the load, keeping the load close to the body and keeping the
back straight while making the lift using the legs. Valuable advice is
given in the Manual Handling Regulations.[17]

Electrical Hazards

Electricity can give rise to electric shock (which can be fatal), burns, falls, fire and explosion. It also gives rise to electric and magnetic fields, whose effects on the body are not yet fully understood.

Fixed wiring

Employers must set up their fixed wiring to adequate standards.[18-20] Insulation prevents access to live conductors. Conductors should be chosen that are adequate for both the intended and the foreseeable fault currents. Devices are available to shut down equipment in the event of a fault (e.g. fuses, residual current devices). Where equipment and supplies have an earth (ground) connection, it is essential that it is connected at all times. Work on electrical equipment must only be carried out by competent persons, who work according to safe practices.[21]

Where a workspace is very large, it is sometimes convenient to connect different areas to different phases of the incoming supply. Where this is the case, it is important to ensure that welders working from different phases do not come into close proximity with one another because this substantially increases the danger.

Electrical equipment

Welding and associated equipment needs to be maintained in a safe condition. Equipment should be inspected to establish that it is in good condition. In the UK there is a requirement to test insulation and earth connections.[22] Inspection can establish that there is no damage to insulation and fittings and that there are no signs of overheating or other faults.

Other items of hand-held electrical equipment, such as grinders, are vulnerable and should be checked formally at relatively frequent intervals (say from three to six months depending on the environment in which they are being used). Larger items, such as the welding sets, which are not moved around frequently, may be formally tested at annual intervals. Testing should be carried out at more frequent intervals if it is apparent that a significant number of faults are being found. This testing does not remove the necessity for the user to make checks regularly, since this is when most potentially dangerous faults are discovered.

Table 1.2. Reference levels for occupational exposure to time-varying electric (E) and magnetic (B) fields (unperturbed rms values)

Frequency range	E-field strength $(V\,m^{-1})$	B field (μT)
<1 Hz	–	2×10^5
1–8 Hz	20 000	$2 \times 10^5/f^2$
8–25 Hz	20 000	$2.5 \times 10^4/f$
0.025–0.82 kHz	500/f	25/f

f is frequency, as indicated in the frequency column.

Electric and Magnetic Fields

Since electric welding processes use very large currents, magnetic fields in the workplace can be larger than those experienced in other occupations. Electric fields associated with welding are low. While most medical studies have shown there is no hazard to health from electromagnetic fields, exposure to large fields can give symptoms, and it is prudent to minimise exposure. There are proposals for the restriction of exposure, given in Table 1.2 above.[23] The unit of magnetic field is the tesla (T). It is difficult to shield magnetic fields, and therefore the normal approach to control of exposure is to avoid entering the area of high field. Typical values from Table 1.2 indicate a suggested limit of 200 mT in a steady field (which is comparable to the field that can be obtained from a good ceramic magnet), falling to 0.5 mT at 50 Hz.

Arc welders should avoid draping the welding cable over their shoulders or wrapping it around their body so as to avoid exposing parts of their body to fields above the guideline figures.

Pacemakers

Pacemakers are medical devices that are implanted in some cardiac patients to regulate their heart rhythm. Electromagnetic fields from welding can affect pacemakers, but the precise effects depend on the type and susceptibility of pacemaker and the cardiac condition that it is intended to correct. Welders who are to have a pacemaker fitted should take advice from their specialist. Most conventional welding equipment will probably not present a great risk. However, equipment that can produce strong pulses of electromagnetic radiation, such as resistance welders may interfere both with the function of the pacemaker and its programming. It is recommended that the areas in the workplace where there is a hazard are labelled to alert pacemaker users to the danger, or that employees and visitors are alerted to the possible danger before they enter the area.

Training

Training is a necessary part of any occupational health and safety programme. The essential elements are:

1 Induction:
    − familiarisation with the workplace
    − welfare arrangements
    − fire and other emergency procedures
    − reporting.
2 Job training:
    − recognition of hazards
    − specific procedures that reduce the risks
    − specific skills that are required
    − hazard warning signs
    − how to recognise when there is a problem, e.g. malfunction of equipment
    − personal protective equipment − when to wear it, how to wear it and how to look after it
    − how to get replacement equipment.

Refresher training should be given and training should be undertaken when the technologies or techniques in the workplace change.

2

First Aid and Accident Reporting

The following is a general definition of first aid:

- In cases where a person will need help from a medical practitioner or nurse, it is treatment for the purpose of preserving life and minimising the consequences of injury and illness until such help is obtained.
- It is treatment of minor injuries which would otherwise receive no treatment or which do not need treatment by a medical practitioner or nurse.

UK Legislation

Legislation[24] requires that employers shall provide equipment and facilities that are adequate and appropriate for the needs of the workplace. If it is decided that trained first aiders are needed, they must have training to an approved standard, along with any appropriate additional training (e.g. for a workplace using cyanide or hydrofluoric acid). In workplaces where very small numbers work, with a very low hazard potential, only an appointed person is required. This person needs the means to summon assistance and to know how to do so.

Approved training providers include:

- St John Ambulance
- St Andrew's Ambulance Association
- British Red Cross Society.

USA Legislation

Where there is no infirmary, clinic or hospital in close proximity that can cater for any injured persons, then the employer must have persons adequately trained to render first aid and the equipment to do so (e.g. facilities for eye irrigation).[25,26] First aid supplies must be maintained, in consultation with a physician. The employer must liaise with the ambulance services.

Suitable training courses are provided by the American Red Cross and others.

General Advice

The standard courses and the manuals which support them[27] cover a wide range of injuries and illnesses. This chapter lists those injuries which are particularly likely in a welding environment; these should be considered in deciding whether any additional equipment, facilities or training should be provided, and discussed with a medical adviser if necessary. The term 'medical aid' is here taken to mean treatment by a doctor at a hospital or surgery.

In any emergency, a potential helper should first consider their own safety. While approaching a casualty, they should observe the surroundings, looking for clues to what has happened and whether there is any danger to themselves. Failure to do this could lead to them becoming an additional casualty.

Electric Shock

First switch off the current, pull out the plug or otherwise remove the casualty from contact with the live conductors. The first aider should not touch the casualty's skin with bare hands until this is done. For normal arc welding voltages, dry clothes will give sufficient insulation. For mains electric supplies, a sturdy non-conducting object such as a wooden broom handle may be used to push the casualty away from the cable.

If the casualty is unconscious, open the airway by tilting the head back and check for breathing. If he or she is still breathing place the casualty in the recovery position and call for medical aid.

If he or she has stopped breathing begin artificial ventilation and check for circulation, giving cardiopulmonary resuscitation if necessary. Call for medical aid.

Where work is done in areas where the risk of electric shock is increased, for example work in damp conditions, extra first aid provision would be advised. Short courses in emergency first aid, including resuscitation techniques, are available.

Burns

Burns can be caused by contact with a live conductor, by arc flash or by contact with hot objects. Burns should be cooled with water for 10 minutes or more, and then covered with a dry sterile dressing. Large or deep burns should be referred for medical aid; a first aider or physician should make this assessment. Electrical or radio frequency (RF) burns should always be referred for medical aid, since there may be internal damage.

Eye Injuries

Foreign bodies and burns

Bottles of sterile eye wash are ideal for washing eyes where dust or grit has entered. However, if anything remains in the eye or is stuck to it, or if the injury has involved a degree of burning, professional medical aid will be needed.

Chemicals

Chemicals in the eye should be immediately washed out with clean water or sterile eye wash. The manufacturer's safety data sheet should be consulted to assess the subsequent actions. If the casualty needs to be taken to hospital or receive further medical aid, it is helpful to take the data sheet with them. In places where operations like etching metals for testing are carried out, or where alkalis or pickling baths are sited, suitable facilities for washing out the eyes should be provided close to the operation.

Arc eye or welder's flash

This is caused by the action of ultraviolet (UV) light on the outer surface of the eyeball. The symptoms develop some time, generally an hour or more after exposure. The eyes are painful, often with a gritty feeling, red, watering and sensitive to light. The eyes should be bathed with cold water and a light covering applied. If there is any doubt about the severity of the injury or a risk that a foreign body may have entered the eye, seek medical aid urgently.

Heat Exhaustion and Heat Stroke

Heat exhaustion and heat stroke are caused by work in a hot environment, often aggravated in the case of welding by the need to wear protective clothing against other hazards.

Heat exhaustion generally develops gradually, and is caused by loss of water and salt from the body. The casualty may be dizzy, confused, with pale clammy skin, and may have cramps in the arms and legs. The pulse is rapid and weak. They should be cooled down and given fluid and salt (one teaspoon per litre of water). If the condition deteriorates, call for medical aid.

Heat stroke is a life threatening emergency, caused by a failure of the body to regulate its temperature. The casualty may have a headache and be dizzy, be restless and confused, hot, flushed and dry. They will have a full pounding pulse. The body temperature of the casualty is greater than normal and rising. Cool the casualty as quickly as possible with water, wet sheets, etc, and get emergency medical aid.

Exposure to Harmful Gases and Fumes

In the context of welding, harmful gases and fumes fall into two main categories, asphyxiating shielding gases and pollutant fumes, as dealt with in Chapters 5 and 6.

Asphyxiating gases displace air, leaving the casualty unable to obtain oxygen. Never enter an area where the atmosphere may be deficient in oxygen without suitable respiratory equipment for which training has been received. More than 50% of those who die in confined spaces are those who are attempting a rescue without proper equipment. If rescue is possible, the casualty should be taken out to fresh air, given resuscitation if required, followed by emergency medical aid.

Vapours arising from solvents used for cleaning, etc, may be asphyxiating and/or toxic, particularly if they are decomposed by UV light. Some decomposition products from chlorinated solvents are highly toxic.

Gases produced in welding can include oxides of nitrogen, and ozone. The former can be produced in large quantities in preheating with a gas torch. They can cause pulmonary oedema, the outcome of which can be fatal. The onset of symptoms is often delayed by many hours and medical aid should be sought promptly if symptoms arise following such activities.

Exposure to excessive amounts of zinc, copper, magnesium and some other metal fumes can cause metal fume fever, with symptoms similar to influenza. Some hours after inhaling the fume the individual complains of tiredness, headache, aching muscles, thirst, a sore throat, a cough, and sometimes of a tight feeling in the chest. The individual will develop a high temperature, have shivering attacks and perspire profusely. Complete recovery normally occurs in 24 to 48 hours. If a worker suffers in this way, and influenza is ruled out, fume levels should be investigated; if no specific pollutant is identified, seek medical advice.

The inhalation of cadmium fumes gives rise to similar symptoms, in some cases followed by acute inflammation of the lungs, which can be fatal. Immediate medical advice should be sought if there is a possibility of illness being due to cadmium fume.

Accident Reporting

Whenever there has been an accident, incident of ill health or a dangerous occurrence that could have led to injury associated with work, it should be recorded and investigated. The primary purposes of such records are to investigate the causes, in order to prevent a recurrence, and to satisfy the authorities.

Certain types of accident, cases of ill health and dangerous occurrences are reportable to the authorities. In the UK the requirements are laid out in the Reporting of Injuries, Diseases and Dangerous Occurrences Regulations, 1995.[28] In the USA the requirements are described in 29 CFR 1904.[29] Employers with more than 10 employees are required to keep a log (OSHA 200, soon to be replaced with OSHA 300) of their accidents and to make certain information accessible to the workforce.

3

Fire

Fire can cause loss of life and destruction of property. Most welding and cutting processes generate significant quantities of heat and many introduce sources of ignition and fuels into the workplace. Therefore an important element of planning most welding operations is the control of the risk of fire. Common causes of fire include electrical faults, sparks, spatter, overheating of combustible materials (e.g. cardboard boxes stored too close to an electric light bulb) and arson.

For a fire to start three essentials must be present:

− combustible material
− oxygen or air
− a source of ignition.

Combustible Materials

Combustible materials include:

1 fuels;
2 wood, often found as blocks to support or wedge work;
3 glass reinforced plastic, which although not normally considered flammable may be ignited by an oxyacetylene flame;
4 cardboard, rags or other packing material;
5 oil or grease;
6 solvents: many non-chlorinated solvents are flammable;
7 dust: flammable dust may accumulate in ventilation systems and once ignited the air flow will fan and distribute the flames;
8 insulation materials, especially foam plastics.

Air

In some welding and cutting operations, particularly gas cutting when the ventilation is poor, the oxygen content of the air may be increased thus increasing the risk of fire. If it is allowed to increase significantly above the normal level, for example to 23%, several problems arise:

– Items that are not flammable in the normal atmosphere can become flammable.
– Fires can be extremely difficult to extinguish.

Thus it is important to ensure that the oxygen level does not rise.

Sources of Ignition

Welding and associated activities such as grinding present constant sources of ignition. Preventing these sources of ignition from having access to combustible material is the key to fire prevention. Matches and cigarette lighters can be ignited by spatter, or the butane gas from the lighter can leak into clothing putting the owner at risk of having their clothing catch fire.

Legislative Framework

In both the UK and the USA, there are a number of statutory obligations relating to fire precautions and the protection of personnel. These should be consulted to ensure that the measures are implemented.

Legislative requirements for the UK are in the Fire Precautions Act, 1971, and the subsequent regulations and guidance.[30–32] Many premises require a fire certificate, which can only be gained when the fire precautions are adequate. Specific items that are assessed before issuing such a certificate are that there are adequate means of escape, that these can be safely and effectively used, that there are means to fight fire and reasonable means of giving warning. All premises must provide adequate means of escape and means of fighting fire. In all cases the provisions must be determined as the result of a risk assessment relating to the property and the activities within it.

Legislative requirements for the USA are found in the general standard for fire protection. There are specific requirements relating to the use of cutting and welding processes.[33–38]

The control of risk from fire generally has several elements: to prevent fire from breaking out, to limit its spread, to raise the alarm and to allow the safe evacuation of all personnel.

Prevention of Fire

The primary means of preventing fire is to keep sources of ignition separated from combustible materials. In particular no smoking should be allowed:

– when handling or connecting oxygen or fuel gas cylinders,
– in the no-smoking zone around the gas cylinder store,
– when using solvents. Note that even if the solvent is not flammable, it will probably be decomposed by heat to give toxic fumes.

Areas where welding is normally conducted should be kept free from all combustible materials. When it is necessary to undertake welding work in areas not specifically designed for the purpose, movable combustible materials should be relocated to a safe area. If any are immovable, safeguards should be used to protect them from heat, sparks and slag. Wooden floors can be wetted or covered with damp sand, sheet metal or equivalent (but provision should then be made to avoid the danger of electric shock). Any cracks or openings should be covered to prevent sparks going through. Blankets, made from material resistant to molten metals, are available; see Fig. 3.1.

Areas that deserve particular attention include:

– combustible materials within about 11 m of the welding,
– openings in that range that expose combustible materials,
– metal walls or pipes adjacent to combustible materials,
– the opposite sides of tanks, or bulkheads, where conducted heat could set fire to materials.

It should be noted that the sparks from air carbon arc cutting and plasma arc cutting can travel up to about 15 m. When work is planned in an area where combustible materials cannot be removed, the use of a 'hot-work' permit is advised. Figure 3.2 shows a typical permit, which specifies the precautions to be taken before work, the permitted duration of work and the precautions to be taken during and after work.

3.1 A blanket that can withstand molten metal (photograph courtesy Tusker, Safety First Manufacturing Co.).

Work on magnesium causes a specific fire hazard. The swarf is especially flammable and may ignite spontaneously in the presence of oil. Once ignited, magnesium is difficult to extinguish. Normal extinguishers are not suitable for magnesium fires; it is essential to get the correct extinguisher and train employees to use it.

Preventing the Spread of Fire

Most large buildings are compartmentalised to delay the spread of fire. These areas will be bounded by fire-resistant doors, which should always be kept closed, unless they are of the type specifically designed to be held normally open and released automatically on triggering the fire alarm. Roof vents which open automatically in the event of fire below them are designed to release smoke before it can spread and damage the rest of the building and thus prevent it from obscuring a firefighter's view, allowing use of the minimum effective amount of water.

Fire can spread readily through ventilation trunking from one space to another. To overcome this, cutoff flaps or fire dampers which are released by temperature rise may be fitted. In many installations it will be desirable to cut off the ventilation fans, but some buildings may be designed to use air pressure to blow smoke out of hazardous areas and such systems must be left running.

HOT WORK
PERMIT-TO-WORK

Building ....   Floor ....

Nature of job (including exact location)

....

The above location has been examined, and the precautions listed on the reverse side have been taken.

Date ...... Time of issue of permit ....
Time of expiry of permit ....

Signature of person issuing permit ....
Signature of person to whom permit is issued ....

NB It is not desirable to issue hot work permits for protracted periods; for example, fresh permits should be issued where work carries on from day to day.

FINAL CHECK-UP

Work area and all adjacent areas to which sparks and heat might have spread (such as floors above and below and on opposite sides of walls) were inspected continuously for at least one hour after the work was completed and were found fire safe.

Signature of employee carrying out fire watch ....

After signing, return permit to person who issued it.

PRECAUTIONS

(The person carrying out this check should tick as appropriate)
Where sprinklers are installed that these are operative.
Cutting and welding equipment is in good repair and adequately secured.

PRECAUTIONS WITHIN 15 m OF WORK
Floor swept clean of combustible materials.
Combustible floors protected by wetting down and covering with damp sand or sheets of non-combustible material.
Combustible materials and flammable liquids protected with non-combustible curtains or sheets.
All wall and floor openings covered with sheets of non-combustible material.
All gaps in walls and floors through which sparks could pass covered with sheets of non-combustible material.

WORK ON WALLS OR CEILINGS
Combustible constructions protected by non-combustible curtains or sheets.
Combustibles moved away from opposite side and clear of any metal likely to conduct heat. (Where metal beams are being worked on, and extend through walls or partitions, precautions must be taken on the far side of such a wall).

WORK ON ENCLOSED EQUIPMENT (tanks, containers, ducts, dust collectors, etc.)
Equipment cleaned of all combustibles.
Containers free of flammable vapours.

FIRE WATCH
Provision for the attendance of an employee during and for one hour after completion of the work. Such employee being supplied with extinguishers or small bore hose and trained in the use of such equipment and in sounding alarm.
Signature of person carrying out the above check ....

3.2   Example of a hot-work permit.

Fire Detection

Automatic fire detection usually senses smoke or heat and it can be difficult to set this to avoid false alarms in the immediate welding environment. However, other areas of the workshop may be covered by such systems. Automatic systems can be set to trigger sprinklers and/or to call the fire service automatically.

The welder will not generally be aware of a fire while arc welding, as it is not visible through the filter glass. While welding in an area which is purpose-built for welding, there should be no risk, since the area will be free from combustible materials. However, where welding is carried out in an area that is not purpose-built for welding, a fire watcher should be on duty. This will be one of the stipulations of the hot work permit. If a fire breaks out the watcher must interrupt the welder and take appropriate action. The watcher should be trained in the use of fire extinguishing equipment and in the means for raising the alarm. After welding has finished, the area should be checked approximately half an hour later, to check that there is no further risk of fire breaking out.

Evacuation and Means of Escape

In the event of an emergency, all persons at risk need to be alerted so that they can go to a place of safety. Ideally, a fire alarm should be readily audible above the noisiest operation and so distinctive that it cannot be mistaken for any other signal. In practice this may be impossible to achieve. It will be necessary to warn individually those workers using particularly noisy processes, by the use of trained fire wardens, for example.

Emergency exits are required where there are not enough normal exits to ensure a speedy evacuation of a building wherever a fire may break out. Many designs of door catches are now available for exits that do not compromise security. Escape routes must be adequately signed, kept clear of rubbish, etc, to allow free exit even in dense smoke, and provided with emergency lighting if necessary.

An emergency plan will need to take account of those who may have extra difficulty in reaching the exit, for example disabled workers, visitors or overhead crane drivers. It is essential to be able to check by assembly and roll call, or by a team of fire wardens sweeping through the premises on their way out to ensure that everyone is clear of the building.

In many instances of fire in the welding shop, the electricity supply can be switched off at the incoming mains, cutting off all arcs and thereby alerting all workers and avoiding the risk of electric shock for those fighting the fire. However, this should only be done if it does not put anyone at risk – for example trapping someone in a lift, crane jib, etc, leaving any equipment in a dangerous state or cutting off lighting needed to escape. Therefore this should receive careful attention when planning the fire procedures.

Extinguishing Fires

Extinguishers rely on cooling the burning material and/or excluding oxygen from it. Water or sand may also be used from a bucket. In general fire extinguishers can only successfully tackle a very small fire. Once established burning has been reached it may be beyond the capability of portable fire extinguishers. It is essential to train staff so that they know how to use extinguishers effectively and how to avoid putting themselves in danger.

Fires involving different materials require different treatment. They have been classified to aid identification and selection of suitable fire-fighting media. However, there are minor differences between the classification systems in the UK and the USA. Table 3.1 gives the classification in the UK.

Fire extinguishers are available for each class of fire and many are suited to more than one class. All new portable extinguishers in the UK are red, with up to 5% of the body (often the writing) in the distinguishing colour as shown in Table 3.2.[39]

The USA classification system for fires is given in Table 3.3.[40] In addition to the coloured marking and lettering, both in the UK and

Table 3.1. UK classification of fires

Class	Description	Typical examples
A	Solid materials, in which combustion takes place with the formation of glowing embers	Wood, paper, rags, paint film, electrical insulation materials
B	Flammable liquids	Solvents, oils or greases
C	Flammable gases	Acetylene, liquefied petroleum gases (LPG)
D	Metals	Magnesium, aluminium powder

Table 3.2. UK identification of fire extinguishers

Type of agent	Colour of extinguisher body	Distinguishing colour	Suitable for use on these classes of fire
Water	Red	White	A (not electric). *Dangerous* to use on classes B and D
Foam	Red	Pale cream	B
Powder	Red	French blue	B, electrical fires (some types can also be used on A or D)
Carbon dioxide	Red	Black	B, electrical fires, small A fires
Halon[a]	Emerald green	–	B, may also be used for A

[a] Halon extinguishers are no longer supplied, but those in service need not be withdrawn until they become due for pressure test.

Table 3.3. USA classification of fires and identification of fire extinguishers[40]

Class	Description of class	Identification of extinguisher (the colours are not mandatory)
A	Ordinary combustibles, e.g. wood, paper, cloth	Green triangle containing a letter 'A'
B	Liquids, greases and gases[a]	Red square containing a letter 'B'
C	Energised electrical equipment	Blue circle containing a letter 'C'
D	Metal fires	Yellow five-pointed star containing a letter 'D'. Also marking to indicate the type of metal.

[a] Fires involving flammable gases are best extinguished by cutting off the gas supply. If the fire is extinguished by other means, the escaping gas may accumulate, then re-ignition (from a spark, hot surface, etc) will cause a substantial explosion.

the USA, pictograms are frequently used to indicate the suitability of extinguishers for particular classes of fire. Extinguishers that are too large to be carried may be mounted on trolleys that can be wheeled to the fire, but hand sizes as well will be needed to reach less accessible places.

Fixed fire extinguishing installations

These are frequently sprinkler systems. Multi-storey buildings may have rising mains, a pipe into which the fire brigade can pump water at ground level drawing it off at whatever level it is needed or hose reels connected to either the normal water supply or to the rising main.

Fire Procedure

The generic fire procedure in the event of fire breaking out is as follows:

- Ensure that the alarm is raised first.
- Leave immediately and go to a place of safety.
- Only attempt to fight a fire if:
    It does not appear beyond control;
    The escape route is clear of fire and smoke.
- Choose the right extinguisher.
- Use it correctly.

In any workplace all personnel should be trained so that they know:

- the location of emergency exits
- the location of extinguishers
- the types of extinguishers to be used for any likely fire
- the operational procedure for the extinguisher (push or pull knob, upright or inverted use, etc),
- the correct application of the extinguisher (direct jet at or near base of fire, etc), if possible by practice.

Ensure that appropriate extinguishers are available beside any especially risky operation. All personnel should receive instruction in the use of all the available fire extinguishers; it is too late to read the instructions when the fire is well established!

Once the fire appears to have been extinguished, keep a close watch on the area until all possibility of re-ignition has been excluded. This will take several minutes. Ensure that used extinguishers are promptly refilled or replaced.

Maintenance of Fire Precautions

The employer should test the alarm systems regularly and keep continual watch on housekeeping standards to ensure that fire exits and fire doors are not obstructed, and that unnecessary amounts of combustible material do not accumulate. Fire extinguishers must be subject to effective preventive maintenance. Any that have a pressure gauge fitted should be checked regularly to see that the pressure is maintained. The employer should set up a planned preventative maintenance scheme to ensure that all fire-fighting equipment is kept in working order.

Regular practice fire alerts should be part of the emergency plan, as they can highlight any deficiencies in the engineering measures, as well as any need to give additional training for staff.

Fire Brigade

The fire brigade should always be notified of a fire immediately, however small it appears at first. The public and/or works fire brigade will also be able to advise on the following items on which they should be fully informed:

- fire prevention measures,
- layout of works to reduce fire risks,
- location of gas cylinder stores, bulk storage tanks, radioactive materials and other special hazards,
- notification of fire, e.g. access to telephone,

The fire brigade should be kept informed of how to identify the site entrance, and where the works are located if the site is large. They should be informed of the layout of the buildings within the site and the hazard areas (such as flammable gas stores and radio-isotopes) by taking a representative round the plant after any major alteration or in any event at, say, two year intervals.

4

Compressed and Liquefied Gases

A number of gases are used in welding and cutting applications. They are used for several purposes. Fuels are used to provide sources of heat for welding and cutting, and these include acetylene, hydrogen, propane and LPG such as butane and MAPP® (methyl acetylene propadiene). These fuels are burned with air or oxygen. Inert or unreactive gases, such as argon, helium, carbon dioxide and many proprietary mixtures are used to provide shielding in the processes that do not use a flux for this purpose. Cutting processes use gases to react with the metal and raise it to its melting point, or as assist gases. Many lasers require a gas supply for the laser itself, or for shielding purposes or as assist gases in laser processing.

With the exception of acetylene, propane and LPG, gases are supplied at pressures up to 300 bar (4300 lb in^{-2}). Gas cylinders are thus pressure vessels and must be constructed and maintained to high standards. They require periodic examination and testing, which must be carried out by the owner of the cylinder; in most cases this would be the gas supplier.

Hazards

The hazards presented by compressed gases include:

- explosion and fire;
- uncontrolled release of pressure – rocket propulsion of the cylinder, injuries to eyes, hearing, gas injection into the blood stream;
- toxicity, asphyxiation, flammability or oxidising action of the contents;
- manual handling injuries.

26

Common Gases used in Welding and Cutting

Acetylene

Acetylene is supplied at a pressure of 15 bar (225 lb in^{-2}) at 15 °C. Acetylene has a distinctive garlic-like odour, is lighter than air, and can therefore collect in roof spaces. Acetylene is highly reactive and may explode if compressed alone; its ignition limits are from 2.5 to 100%. It can only be stored safely in cylinders specifically designed for acetylene service. Cylinders are filled with a porous substance, such as charcoal, kapok or kieselguhr, which is soaked with acetone, see Fig. 4.1.

The gas dissolves safely in the acetone. Hence cylinders of acetylene are sometimes known as 'dissolved acetylene' or 'DA'. The porous mass is designed to slow down or prevent the gas from decomposing. The time from initiation to explosion should be several hours. However, it can be much quicker if the mass inside the cylinder has been damaged, say by repeated flashback, by mishandling of the cylinder such as dropping, if the valve is leaking or if the pressures are too high. Mechanical shock to the cylinder, or overheating, may cause decomposition of the gas inside and this may lead to high temperatures inside with possible detonation. Acetylene can also react with copper and silver, and alloys containing these in sufficient quantity, to form explosive acetylides.

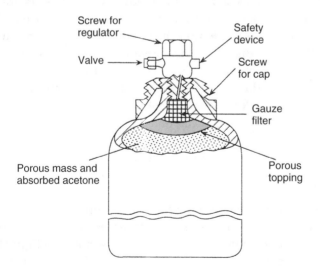

4.1   Acetylene gas cylinder.

Acetylene cylinders must always be stored and used in an upright position and the gas should not be withdrawn at a rate exceeding one fifth of the cylinder contents per hour, to avoid withdrawing acetone with the acetylene. Only acetylene regulators should be used.

Argon, helium, nitrogen and carbon dioxide

Argon, helium and nitrogen are non-toxic, but do not support life. Therefore cylinders of these gases present the range of hazards associated with pressure release and under certain circumstances can present an asphyxiation hazard. Entry into an area with a high concentration of these gases (and hence a low concentration of oxygen) can cause almost immediate unconsciousness, without any warning signs being noticed by the victim. Exposure to excessive carbon dioxide levels can cause headache, dizziness, tinnitus and raised respiratory rate with difficulty in breathing. Ultimately it can lead to drowsiness and unconsciousness. The compound is supplied as a liquid in cylinders which are fitted with a dip tube; a vaporizer changes the liquid into a gas before it passes to the regulator. This group of gases is used principally for shielding.

Hydrogen

Hydrogen is chiefly used for heating and cutting. It is lighter than air, colourless, odourless and non-toxic. It is an explosion hazard and is extremely flammable. Concentrations between approximately 4% and 75% will burn with a flame that is almost invisible. Special care should be taken to use only hydrogen regulators. The same thread is used for the other fuel gases, LPG and acetylene, and so regulators designed for use at a much lower pressure with these gases could be connected accidentally to hydrogen cylinders if personnel are not trained to check, or if care is not taken.

LPG and propane

This group of gases include several gases and mixtures, such as butane and MAPP® (methyl acetylene propadiene mixture). Propane is supplied in large diameter red cylinders and stored in liquid form. A propane regulator should always be used. The cylinders must always be stored and used in an upright position to avoid the

to provide protection
ps or spray if the store
ls used to construct the

vicinity of the fuel gas
o avoid ignition of any
on, no smoking should
sign should be erected

and secured in, racks
ether full or empty. As
nose for acetylene and
ack should not be more
ylinders at the bottom.
ne and propane must
osition. Other products
ecially oils, paints and

both the gas supplier's
on transport. The area
and be kept clean and
responsible storeman,
take in the event of an
tore should indicate the
ithin and the name and

be protected to prevent
as have a screw-on cap,
ed to the cylinder; this
aky valve cannot accu-
ld not normally be used

suppliers and from the
create a store inside a
gregation distances.

ly dangerous 'gases' can
orage tanks in the user's
pplier's road tanker.[50–52]

discharge of liquid instead of gas. LPG and propane are heavier than air and can collect in drains and trenches. Vapours can flow along the ground and can be ignited at some distance from the source. They present potential fire, explosion and asphyxiation hazards. Propane is frequently stenched.

Oxygen

Oxygen is crucial to life, but if present in excess can create a significant fire hazard. Most metals, especially in powdered form, burn in oxygen. Fires in an oxygen-enriched atmosphere are extremely difficult to extinguish. Oils, greases and other organic materials can spontaneously explode on contact with pure oxygen. Only oxygen regulators should be used. Pipelines for oxygen service must be thoroughly cleaned before use. Oxygen should never be used for driving pneumatic tools, inflating vehicle tyres, cooling or refreshing air in confined spaces, or for dusting down apparatus – it is not a substitute for air.

Identification of Cylinders

Gas cylinders are painted in distinctive colours[41,42] to aid identification. As a precaution against connection of the wrong regulator, the outlet threads of fuel gases are left hand, whereas oxygen and inert gases are right-hand threaded. Examples of common gases used in welding and cutting are given in Table 4.1. However, the *primary* means of identification of the contents of the cylinder is by the label which is attached to the shoulder of the cylinder and this should always be checked. If the label is unreadable, the cylinder should be returned to the supplier.

Table 4.1. Examples of typical gas cylinders (UK)

Gas	Cylinder colour	Thread on outlet
Acetylene	Maroon	Left hand
Hydrogen	Red	Left hand
LPG	Red	Left hand
Oxygen	Black	Right hand
Helium	Brown	Right hand
Argon	Blue	Right hand

While the outlets are left- or right-hand threa
ing the wrong regulator, the user must still
to ensure that the regulator is suited to the
regulator or other equipment not rated for th
tents of the cylinder can result in catastrophic
and fatal accidents are possible.

Manifolded Cylinders

If a greater delivery rate of acetylene or a larg
is required than can be obtained from a single
cylinders could be manifolded together. The p
subjected to full cylinder pressure, so it is e
made from materials completely compatible
tested before use. Some gases may be purchas
ders (Fig. 4.2), permanently assembled in a
to a single outlet. The purpose of the flashback
in Chapter 10.

4.2   Cylinder 'bundle' with flashback ar
supplying a piped gas installation (court

roofing, again of lightweight construct
against rain if the store is outside, or o
is within an enclosed workshop. The ma
shelter should be non-combustible.

Any electrical fitting within the imme
store should be of flame-proof construct
accidental escape of gas.[49] For the same
be allowed in or near the store. A promi
to remind people of the prohibition.

Cylinders should be easily placed in
clearly marked to indicate gas type and
an alternative, gas cylinders, other tha
propane, may be stacked horizontally. Th
than four cylinders deep, with the larg
Cylinders must be safely wedged. Ace
always be stored and used in the vertica
must not be left in the cylinder store,
corrosive liquids.

Adequate access should be available
delivery transport and the user's distrib
should be used for storage of cylinders o
tidy; this can best be done by appointin
who can then be trained in what action
emergency. A permanent notice outside th
type and location of all the gas cylinders
location of the store person.

When not in use the cylinder heads sho
damage to the valves. Some cylinder des
but many have a shield permanently sec
demands no user action and gas from a
mulate inside, see Fig. 4.3. This shield sho
for lifting the cylinder.

Advice should be sought from the ga
enforcement authorities if it is intended
building. There are strict rules regarding

Bulk Storage

Oxygen, propane and several other potenti
be stored in bulk in liquid form in special
works; the tanks are refilled from the gas s

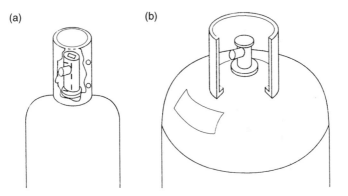

4.3   Gas cylinders with a permanent valve shield. (a) Hydrogen with shield partly cut away; (b) propane.

Acetylene cannot be supplied in this way and manifolded cylinders must be used. The distribution pipelines must meet stringent standards regarding leakage, protection from damage and fittings, such as flashback arrestors.[53-55] Acetylene pipelines must also be free from pure copper and silver, either in the pipe itself or in silver solder for fittings.[56]

The safety precautions needed in the installation of both storage tanks and pipelines are usually met by entrusting the work to a specialist firm, responsible to the gas supplier, who will need to be assured of its suitability before commencing bulk supply. Large scale oxygen supplies may be 'odorised' or 'stenched' to assist in the detection of leaks which could otherwise cause oxygen enrichment.

Bulk storage and pipeline distribution reduce the handling required for cylinders, improving safety and reducing costs; in fact, bulk storage may be economic even at low rates of use.

Transporting Cylinders

Manifolded cylinders can be readily moved by a forklift truck or a crane. Employers should check that their vehicles and lifting equipment are suitable, and in particular that the forks of their fork lift trucks are sufficiently long. Gas cylinders must be treated with care and not subjected to mechanical damage, falls or undue heating. If cylinders have to be handled by means of a crane they should be secured in a special carrier. On no account should an ordinary chain sling, rope or sling, or a magnetic lifting device be employed. Where

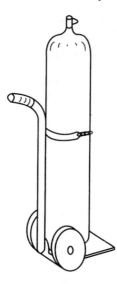

4.4   Trolley to move cylinders.

cylinders are moved with a fork lift truck, they must be secured so as to avoid rolling off the forks. The valves, too, will be exposed to risk of damage if narrow doorways have to be negotiated and special care should be taken in such circumstances.

Care is also needed if cylinders are moved by hand. To avoid damage to the cylinder, or injury or strain to the person moving them, it is best to use a trolley. There is guidance for the avoidance of manual handling injuries issued by the British Compressed Gases Association.[57] Cylinders must not be dragged or rolled along the ground. A serious fire was caused by an employee dragging a cylinder along the ground – the sparks from the abrasion of the steel against the concrete ignited a cloud of flammable vapour. Cylinders must not be lifted by the valve or valve protection cap unless it has been specifically designed for that purpose. Where portable plant is required, the oxygen and fuel gas cylinders should always be transported on a suitable trolley, which should never be allowed to rest horizontally, see Fig. 4.4.

When handling the gas cylinders, personnel should wear protective footwear[58] and industrial gloves. Loose clothing, especially loose sleeves, which can catch on the cylinder valve, should be avoided. 'Milk churning' cylinders is permissible, but it requires practice and should not be used for long distances or on uneven ground. Employees should be trained not to try to catch a cylinder that is toppling, but to let it go. Cylinders should *never* be transported with the regulator attached, unless they are in a purpose built trolley, and in this case the valves *must* be shut.

Cylinder Valves

Some cylinder valves have handwheels, others have a valve with a square end on the stem and must be operated by the correct key to avoid damage. Keys should always be at hand where such cylinders are in use, so that a cylinder can be closed quickly in an emergency. If a cylinder valve is found to be leaking and cannot be closed with the application of moderate force, the cylinder should be placed where the leaking gas will disperse safely and then returned to the suppliers, advising them of the problem. If there is a leak from the valve spindle, it is permissible to tighten the valve nut to a moderate extent with a spanner.

Valves are opened by turning anticlockwise and closed by turning clockwise. They should never be wrenched shut, but only turned sufficiently to ensure that the gas flow is cut off. When opening, the valve should not be left in the fully open position, because it may seize. It should be turned back half a turn. An acetylene valve should not be opened by more than three revolutions.

Security

Because oxyacetylene equipment is readily portable, it is attractive to thieves and vandals; adequate security measures should be taken. The arrangements should be such that sets are not locked in unventilated storerooms or cupboards overnight. Any gas leak carries a serious risk of fire, and while they are more secure locked in such cupboards, they are a big fire risk. The local electrical equipment is unlikely to be adapted to flammable atmospheres, and the emergency services will be unable to identify and locate the cylinders quickly.

Regulators and Gauges

Regulators must always be fitted to the cylinders to reduce the gas pressure from the cylinder pressure to the working pressure of the blowpipe. Only regulators of an approved standard[59-62] and designed for the gas being used may be fitted to the cylinders. Simple needle valves are not permissible because they will not prevent pressure in the blowpipe and hose lines rising each time the control valves are closed or if the nozzle becomes blocked, nor will they prevent a reverse flow of gas towards the cylinder.

Before a regulator is fitted, the threads and the seats of the cylinder and the regulator should be inspected. If either is damaged, it may be difficult to obtain a leak-tight seal. The outlet to the cylinder must be clean and dry. If it does not seal properly, it should be dismantled, cleaned and tried again.

It is common practice to 'snift' or 'crack' cylinders to clean the outlet. This practice consists of briefly opening the main valve of the cylinder and quickly closing it again. It is a potentially dangerous practice, since the gas emerges at very high speeds. On no account should a hand or other part of the body be placed in the gas stream. Eye protection should be worn and care should be taken that there is no source of ignition nearby. If snifting is attempted on hydrogen, the emerging gas may ignite spontaneously. Thus this practice should never be carried out with hydrogen. It is recommended that snifting is not used, but that the outlet is cleaned by wiping and/or the use of a low pressure compressed air jet. Never try to lubricate the threads, or to use tape to seal the joint. If the regulator does not screw in easily, then do not force it – it is probably the wrong type.

The adjusting screw of the regulator must always be turned fully anticlockwise before the cylinder valve is opened, which should be done slowly. The regulators should always be detached from the gas cylinders before moving them, unless they are being conveyed on a trolley or other special truck.

Pressure gauges are generally an integral part of the pressure regulator. Only use gauges supplied for the purpose.

Planning for Emergencies

Excessive heat will cause an increase in pressure inside the cylinder. This may result in the cylinder bulging or even exploding. Allowing an arc to play on a cylinder wall may have the same effect.

When planning the cylinder store it is appropriate to consider whether a trough and water supply can be installed at a suitable distance to facilitate the emergency cooling of a heated acetylene cylinder. Fire extinguishers or other fire protection or suppression systems or devices should be available where flammable gas cylinders are stored – CO_2 or chemical types. The fire brigade should be kept informed of the store location, which should be clearly marked.

If, in spite of all precautions, fire does break out in the gas store, the cylinders may eventually explode on prolonged heating, so an emergency plan should be drafted, and communicated to all staff. Key staff will need to be trained. The fire brigade should be alerted immediately. The area around the gas store should immediately be evacuated to a distance of at least 100 m. Personnel between 100 m and 300 m away should take cover. The emergency team should remove cylinders not immediately involved in the fire, and which have not been heated, to a safe place if it is possible without taking personal risk. Burning gas should not be extinguished unless the gas flow can be cut off immediately. It is safer to burn flammable gas than to risk building up an explosive mixture.

Cylinders that have been involved in a fire must be cooled. They should be sprayed with water from a safe position until the steaming from the surface stops. They will then need to be removed to a safe place and the supplier notified so that they may be disposed of safely. This measure is not sufficient in the case of acetylene cylinders, owing to the unstable nature of the gas. Acetylene cylinders must be cooled until they remain wet. Only then may they be approached so that the surface of the cylinder may be felt. If it remains cold for at least one hour then the cylinder may be submerged in water, where it should be left for at least 12 hours before it is removed for disposal.

5

Fume, Dust, Vapour and Gases

This chapter deals with the nature of fume, its origins and the harm that it can do. It also outlines the exposure limits. Methods of control are discussed in Chapter 6.

Fume, dust, vapour and gases can all be inhaled. The harm that they can cause depends on:

- their chemical nature
- their particle size (in the case of dusts and fume)
- their solubility
- the quantity absorbed
- the duration and frequency of exposure
- the occupational environment
- the susceptibility of the individual.

Particle Size and Behaviour

Dusts and fume are suspensions of finely divided solid particles in the air. Their diameter determines how long they remain in the air and how far down the respiratory pathway they can reach. Particles above 100 μm diameter fall very quickly and remain near the point of emission. They are unlikely to be inhaled. Particles in the range 30 to 100 μm also fall out quickly but can be carried further by air currents. At the smaller diameter end of the range particles can be inhaled, but will be trapped by the filtering mechanism in the nose. They are unlikely to be absorbed unless they are soluble in water.

In the range 5 to 30 μm, particles remain airborne for some time. On inhalation they will reach the bronchial part of the respiratory system. From there they will be cleared by the ciliary mechanism.

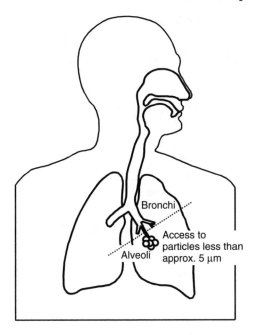

Bronchi

Access to
particles less than
approx. 5 μm

Alveoli

5.1   Areas of the respiratory tract reached by particles.

Particles below 5 μm will be carried into the air sacs in the alveolar region. They are removed only slowly from this region. However, particles that are less than 1 μm in diameter will frequently remain suspended in the air and be expelled with the exhaled air, see Fig. 5.1.

Humans are particularly at risk from the inhalation of substances, first because they cannot stop breathing and second because the quantity of air breathed in a day is very large. Thus even small concentrations of a substance in the atmosphere have the potential to affect the health of the individual. Gases and vapours are inhaled deep into the alveolar region. Their effect depends primarily on their chemical nature.

The welder may be exposed to a wide variety of dusts, fume and gases. Some welding and cutting operations generate large quantities of fume. The size range of the particulates produced spans all those referred to above and can be compared to other airborne particulates, see Fig. 5.2.

In the welding environment, sources of substances that can be a respirable hazard include:

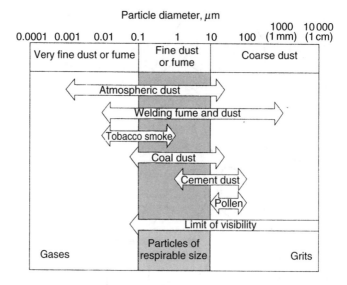

5.2   Sizes of particles.

– Welding
      the parent metal,
      the filler and its flux,
      shielding gases,
      the action of the heat source on the environment,
      surface contaminants;
– Grinding dusts;
– The welding environment itself, which may be deficient in
   oxygen, or contain asbestos, chemical hazards, oils, etc.

Health Effects of Inhaled Substances

The manufacturers of welding consumables and the suppliers of
gases are required to supply data sheets, giving details of the harmful
effects, if any, that the products present. They also generally supply
typical figures representing the approximate composition of the fume
when used under typical conditions. It should be borne in mind
that the data sheet for a substance or preparation may not foresee
all the conditions in the workplace – it may not, for example, give
the breakdown products that are likely to arise close to an arc.

   Even a totally innocuous substance, if inhaled in sufficient quantity, can interfere with the efficiency of the lungs. Many of the substances that are inhaled in a welding environment are irritants, that

is, they produce local swelling and reddening of the surface of the part of the body exposed.

Some substances are toxic when inhaled, producing immediate or chronic effects. They may be absorbed into the blood stream and carried around the body to a target organ such as the liver or the kidneys. Chronic effects, for a few substances, can include cancer.

A few substances can produce sensitisation by inhalation or by skin contact. Substances that cause sensitisation by inhalation give rise to an allergic form of asthma.[63,64] Cases of asthma associated with isocyanates, or with solder fluxes containing colophony, are relatively common. Cases of asthma associated with welding stainless steel are also reported. Skin contact with such substances can give rise to allergic dermatitis.

A number of metals give rise to an illness known as metal fume fever. This manifests itself as a flu-like illness which comes on some hours after exposure, with a fever, aches and pains. Sufferers usually make a full recovery if the cause was a metal such as zinc, but the outlook can be more serious if the cause was exposure to a toxic metal such as cadmium. In case of doubt, medical attention should be sought promptly.

Tables 5.1 and 5.2 give the health effects and exposure limits (correct at December 2000) for some of the common substances encountered in welding.[65-72] Except where indicated, the exposure limits relate to an eight hour time-weighted average.

Some of the substances in the tables above are by-products of the welding operation, whereas some are perhaps encountered in the environment. The following section deals with some of the possible by-products.

By-products of Welding

Nitrogen oxides

One by-product of flame processes, and to a lesser extent arc processes, is that nitrogen from the air reacts with oxygen to form nitric oxide and nitrogen dioxide. Both of these gases are toxic and are a potentially serious hazard if present in excess. Fatal accidents have occurred when preheating steel in areas with insufficient ventilation. The effects of overexposure are not apparent for some hours, the person being exposed is not aware of the danger, and therefore adequate preventive measures are essential.

Table 5.1. Adverse effects and exposure limits relating to inhalable particles[65-72]

Substance	Possible adverse effects	Exposure limit (UK) (mg m^{-3})	Exposure limit (USA) (mg m^{-3})
Aluminium	Long-term inhalation of aluminium powder may cause scarring of the lungs	4 (respirable dust) 5 (welding fume)	5 (respirable)
Antimony	Irritant. Disturbs the metabolism of protein and carbohydrate	–	0.5
Arsenic	Carcinogen, toxic. Target organ is the liver	0.1[a]	0.01, averaged over 8 hours[71]
Asbestos	Causes mesothelioma, lung cancer and asbestosis	0.2 fibres/mL for 4 hours, 0.6 fibres/mL for 10 min (not chrysotile)[69]	0.1 fibre cm^{-3} [70]
Barium	Highly toxic, muscle stimulant	0.5 (soluble dusts)	0.5 (soluble) 5, non-soluble, respirable
Beryllium	Highly toxic. Over-exposure may cause fatal berylliosis (chemical pneumonia)	0.002[a]	0.002[b]
Cadmium	Highly toxic, carcinogen. Long-term exposure causes emphysema. Causes metal fume fever; even small overexposures may cause fatal illness	0.025[a]	0.1[b]
Chromium	Prolonged exposure to chromium metal dust can cause lung fibrosis. Chromium VI compounds are suspected carcinogens	Cr (VI)[a] 0.05 Cr (III) 0.5	Cr (VI) 0.1[b] Cr (III) 0.5
Cobalt	May cause sensitisation by inhalation or skin contact. Moderately toxic	0.1[a]	0.1
Copper	May give rise to metal fume fever. Inhalation may cause muscle weakness and headache. Some copper compounds are highly toxic	0.2	0.1 (fume) 1 as dusts
Fluorides	Toxic by ingestion, highly irritating. Causes pulmonary oedema over a long term. Has a chronic effect on bone	2.5	2.5
Iron	Inhalation over long periods may cause scarring of the lungs without physiological symptoms (known as siderosis)	5 (fume)	10 (iron oxide fume)

Table 5.1. (*continued*)

Substance	Possible adverse effects	Exposure limit (UK) (mg m^{-3})	Exposure limit (USA) (mg m^{-3})
Lead	Fatigue, headache, sleep disturbance, gastric disturbance, anaemia. Attacks the central nervous system	0.15[67]	0.05, averaged over 8 hours[68]
Manganese	Skin and eye irritant. Causes degenerative changes in the brain. Causes metal fume fever	1 (under review)	5 (C)[c]
Mercury	Corrosive to the skin, eyes and mucous membranes. Causes stomach pain, diarrhoea, kidney damage and respiratory failure	0.025	0.1
Mild steel welding fume		5	
Molybdenum	Ill health from occupational exposure is unlikely	5	5 (soluble)
Nickel	Harmful. Repeated skin contact may cause allergic contact dermatitis. A suspected human carcinogen	0.1 (soluble)[a] 0.5 (insoluble)[a]	1
Silver	Inhalation causes argyrosis	0.01	0.01 (soluble)
Tin	Tin powder may cause irritation. Inhalation over a long period may cause scarring of the lungs without physiological symptoms	2	2 (not oxides)
Titanium	Occupational ill health from exposure is unlikely	4 (respirable)	15 total dust
Tungsten	Occupational ingestion is not known to cause long-term ill health	5 (insoluble)	–
Vanadium	Overexposure causes chest pain, cough, difficulty in breathing, asthma, bronchitis	–	0.5 (C)[c] respirable dust 0.1 (C)[c] fume
Zinc	Can cause metal fume fever, but only moderately toxic	5 (zinc oxide fume)	15 total dust 5 respirable

[a] Maximum exposure limits – exposure must be reduced so far as is reasonably practicable, and in no case exceed these limits.
[b] Substances that have ceiling values, from Table Z2 of regulation 29 CFR – consult the regulation 29 CFR 1910.1000 for more details.[72]
[c] (C) is a ceiling value.

Table 5.2. Adverse effects and exposure limits relating to gases and vapours[65,66]

Substance	Common sources	Possible adverse effects	Exposure limit (UK) (ppm)	Exposure limit (USA) (ppm)
Argon, helium	Shielding gases	Asphyxiants	–	–
Carbon dioxide	Shielding gases, combustion of fuels	Asphyxiant	5000	5000
Carbon monoxide	Partial combustion of fuels, decomposition products	Blocks the attachment of oxygen to haemoglobin	30 (long term) 200 (15 min)	50
Oxides of nitrogen (especially nitrogen dioxide)	Action of welding torch on the gases in the air	Pulmonary oedema, which may be fatal. Shortness of breath, coughing, etc	3 (long term) 5 (15 min)	5 (C)[c]
Oxygen	Accidental release	Assists fire	–	–
Ozone	Action of UV on air near the weld	Irritant. In excess causes pulmonary oedema. Thought to have long-term effects on the lungs	0.2 (15 min)	0.1
Phosgene	Action of arc on chlorinated degreasing comounds	Highly toxic, produces hydrogen chloride in the lungs	0.02 (8 hour) 0.06 (15 min)	0.1
Trichloroethylene	Degreasing	Mildly toxic. Produces headache, drowsiness. Overexposure can be fatal	100[a]	100[b] 200 (C)[c]

[a,b,c] See foonotes to Table 5.1.

Ozone

Ozone is formed by the action of ultraviolet light on the oxygen in the air:

$$3O_2 \rightarrow 2O_3$$

Ozone is chemically very reactive and toxic. Contact with most solids will cause it to revert to oxygen. Ozone also attacks rubbers, causing them to perish, so rubber hoses, etc, should not be used where ozone levels are likely to be high. Because ozone so readily reverts to oxygen, the worse is the fume emission from a welding process, the smaller is the amount of ozone likely to be produced. It is claimed that the addition of a small amount (0.05%) of nitric oxide to the argon shielding gas can inhibit its formation. This shielding gas is patented and available commercially. Because it can revert to oxygen by reacting on surfaces, a simple mask can help to protect a welder. Much of the ozone in the atmosphere originates from other sources, for example near fluorescent lamps, and as a result of the action of sunlight on polluted air.

Carbon monoxide

Carbon monoxide is produced as a result of incomplete combustion in a gas flame. Thus it is likely to be more evident in a reducing flame. It may also be produced as one of the decomposition products of fluxes.

Phosgene

Phosgene is a highly toxic chemical that can be produced by the action of ultraviolet light on chlorinated solvents. Hence it can be avoided by the exclusion of all such solvents from anywhere near arc processes.

Composition of Fume in Welding

Because the parent metal is raised to its melting point in welding it is likely to generate fume, usually containing the oxides of the metal. However, the fume may also contain the metal itself, so, for example, welding galvanised steel sheet can give rise to considerable quantities of zinc dust in the fume, to the extent that the fume can be pyrophoric. Table 5.3 lists the probable major constituents in fume when welding alloys. A discussion of the fume in relation to process variables is given in an HSE publication.[73]

Quantity of Fume

While these are the chief constituents of fume, only some are likely to exceed the allowable limits. Table 5.4 gives an indication of the processes and materials that may cause problems. It must be stressed

Table 5.3. Constituents of alloys

Material	Constituents[b]
Steels	
mild steel	Iron
stainless steels	Iron chromium, nickel, molybdenum, cobalt
high-yield steel	Iron, *manganese*
Aluminium alloys	Aluminium, *manganese, zinc* (some contain copper but are not usually welded)
Magnesium alloys	Magnesium, aluminium, *zinc, manganese, thorium*
Alloys with special names	
aluminium bronze	Aluminium, *copper*
brass	*Zinc, copper*
cupronickel	*Nickel copper*
German silver	*Copper, zinc, nickel*
gunmetal	*Copper, zinc, lead*, tin
Incoloy[a]	*Nickel*, iron, *chromium, copper*
Inconel[a]	*Nickel*, iron, *chromium*
manganese bronze	*Copper, manganese*
Monel[a]	*Nickel, copper*
Nimonic[a]	*Nickel, chromium, cobalt*
phosphor bronze	*Copper, lead*, tin
soft solder	*Lead*, tin
Stellite[a]	Cobalt, chromium, tungsten, nickel
Alloys for casting	
aluminium-based alloys	Aluminium and some contain *copper*
copper-based alloys	*Copper* and some contain *lead*

[a] Trade mark or trade name.
[b] Constituents in italics contribute significantly to toxic fume.

that Table 5.4 is not definitive, because good general ventilation is assumed. The employer must assess the process in his own work-place, and gauge its potential.

Coatings

Items to be welded could be plated. This could be chromium, nickel, lead, zinc or cadmium. It is difficult to distinguish between them entirely from their appearance. It is advised that they are removed by machining. Lead plate for fuel tanks is known as 'terne'; this is usually resistance welded in production when fumes may be kept under control by ventilation.

Painted items could contain a number of metals. Yellow paint frequently contains lead. Other paints contain zinc, mercury, arsenic and copper. Anti-fouling paints are a potential problem. Paints

Table 5.4. Significant pollutant constituents. (Constituents with exposure limits higher than the total inhalable dust level, such as iron and aluminium, have been omitted.)

Process and parent metal	Nuisance particulate	Barium	Beryllium	Cadmium	Chromium	Cobalt	Copper	Fluorides	Lead	Manganese	Nickel	Silver	Carbon monoxide	Oxides of nitrogen	Oxygen enrichment	Ozone
Welding																
Oxygas mild steel	1												1	1	1	
Lead	2								3							
Manual metal arc welding (MMA)																
Mild steel	2	(2)					1						1	1		1
Manganese steel	2	(2)					1			2			1	1		1
Stainless steel	2	(2)			3	(2)	1				2		1	1		1
Gas shielded metal arc																
Mild steel: CO₂ gas	2						1						2			1
Mild steel argon gas	2						1						1			1
Stainless steel	2				2						2		1			2
Manganese steel	2									2			1			1
Aluminium	2															2
Copper	2		(3)				3									1
Mild steel, flux cored wire	2	(2)					1		1				1			1
Tungsten inert gas welding (TIG)[a]																2
Mild steel	1			1												2
Stainless steel	1			1						1						2
Aluminium	1															2
Copper	2		(3)			2										2
Nickel	2										2					2
Submerged arc																
Steels	1															
Cutting and gouging																
Oxygas mild steel	2								2[b]				1	2	2	
Air-arc mild steel	2						2		2[b]				1	1		2
Oxy-arc mild steel	2								2[b]				1	1	2	2
Plasma mild steel	2								2[b]				1	1		2
Plasma aluminium	2													1		2
Brazing, all processes	2			(2)								2				
Soldering, all processes	2								1							
Flame preheating													2	2		

1 Significant amounts of fume, but not usually exceeding the limits.
2 Precautions recommended, such as local exhaust ventilation, since exposure levels are generally above the exposure limits.
3 Potential danger from the quantity in relation to its toxicity.
() Occasional occurrence only.
[a] TIG welders may be exposed to thorium particulates when grinding their electrodes.
[b] Lead painted or lead coated steels.

frequently use organic binders which may decompose in the welding process.[74]

Plastic Coatings

Plastic and synthetic resins are now commonly used in coating materials for protection of metal surfaces and as insulating layers between metals. Burning and welding such coated metals invariably gives rise to organic decomposition products, many of which are known to be hazardous to health. All coated surfaces must be treated as potential sources of noxious fumes, particularly where the nature of the coating is not known (as is usually the situation during maintenance and repair work).

If the nature of the coating materials is known it is sometimes possible to predict the decomposition products. In view of the great number of synthetic resins in use and the variety of formulations that are possible, no comprehensive list of air contaminants can be given. Each case must be reviewed separately. Even then the full answer can be obtained only after actual tests on the coating in question. From past experience it is possible to make some generalisations about the air contaminants likely to arise from certain classes of coating material and these are tabulated as Table 5.5: a detailed study has been undertaken.[75]

In addition to the organic materials mentioned above, the coating material may contain a variety of metals and inorganic materials such as zinc, zinc chromate, iron oxide, aluminium, etc, all of which may create extra hazards. The complexity of modern coating materials is such that several different resins may be used in the same formulation, as well as novel materials not mentioned above.

The quantity of noxious compound evolved is dependent on a number of factors, including:

- formulation of coating;
- area of coating being burnt;
- thickness of coating;
- temperature of pyrolysis;
- presence of oxygen excess or deficiency.

The last two factors may also affect the nature of the pyrolysis products: for example the absence of oxygen will favour the formation of carbon monoxide and unsaturated compounds. It should be remembered that, during welding, the reverse side of the metal will also be

Table 5.5. Contaminants likely to be formed by heating paint and plastics coatings

Elements present in resin	Chemical classification of resin	Possible products of pyrolysis
Carbon, hydrogen and possibly oxygen	Resin and derivatives Natural drying oils Cellulose derivatives Alkyd resins Epoxy-resins (uncured) Phenol–formaldehyde resins Polystyrene Acrylic resins Natural and synthetic rubbers	Carbon monoxide Aldehydes (particularly formaldehyde, acrolein and unsaturated aldehydes) carboxylic acids Phenols Unsaturated hydrocarbons Monomers, e.g. from polystyrene and acrylic resins
Carbon, hydrogen, nitrogen and possibly oxygen	Amine-cured epoxy resins Melamine resins Urea–formaldehyde resins Polyvinyl pyridine or pyrrolidine Polyamides Isocyanate (polyurethanes) Nitrocellulose derivatives	As above, but also various nitrogen-containing compounds, including nitrogen oxides, hydrogen cyanide, isocyanates
Carbon, hydrogen and possibly halogens, sulphur and nitrogen	Polyvinyl halides Halogenated rubbers PTFE and other fluorinated polymers Thiourea derivatives Sulphonamide resins Sulphochlorinated compounds	As above, but also halogenated compounds. These may be particularly toxic when fluorine is present Hydrogen halides Carbonyl chloride (phosgene) Hydrogen sulphide Sulphur dioxide

raised to a high temperature, which may give rise to fumes in other compartments or rooms.

The increasing use of plastic foams or expanded plastics, e.g. polystyrene or polyurethane foams, for thermal or acoustic insulation can present an additional hazard during maintenance and breaking up operations. Because of the presence of large amounts of organic material, cutting or burning the insulated section could produce dangerous concentrations of air contaminants.

Asbestos

Welders are among the groups of employees, which includes maintenance workers, shop fitters, etc, who may be called upon to dismantle or alter a structure where there is asbestos cladding present. They should be taught to recognise asbestos and to take appropriate action. The removal of asbestos is a specialised operation, because it is important to avoid the spread of loose fibres into the environment.

Exposure Limits

Tables 5.1 and 5.2 list the exposure limits of some of the more common constituents of fume. In the UK, exposure to substances harmful to health is governed by the Control of Substances Hazardous to Health Regulations (COSHH),[76] with the exception of lead and asbestos, which have their own regulations.[67,69] COSHH requires that the exposure to such substances is controlled within the exposure limits. In particular, for a relatively small number of substances, a maximum exposure limit is set. These are designated by [a] in Tables 5.1 and 5.2. For these substances, exposure must be controlled to the lowest level reasonably practicable and exposure must never exceed the limit. For the remainder of the substances, control of exposure is deemed to be adequate if it remains below the exposure limit. Occasional excursions are acceptable provided that the employer takes steps to remedy the situation.

In the USA, the exposure to air contaminants is governed by standard 29 CFR 1910.1000.[72] Ceiling values in Table Z1 of the standard must not be exceeded, and exposures must be kept within the time-averaged limits. In prescribed cases, ceiling values in Table Z2 of the standard may be exceeded.

To keep substances below the exposure limits, engineering controls must be determined and implemented wherever feasible, along with administrative controls. If these do not achieve full compliance, then personal protective equipment should be used to achieve compliance. These measures are described in the following chapter.

6

Control of Exposure to Fume, Dust, Vapour and Gases

Air

Clean, dry air is approximately:

- Oxygen: 20.94% by volume
- Carbon dioxide: 0.03%
- Nitrogen and other inert gases: 79.03%.

In an indoor workplace, there must be a flow of air to dilute contaminants, to remove excess heat and to maintain comfort. A human being requires approximately 0.15 litres per second (ls^{-1}) when sedentary, and up to around $1.0 ls^{-1}$ when undertaking heavy work. Approximately $2 ls^{-1}$ of fresh air is needed to dilute the exhaled carbon dioxide.

Control of Exposure to Fume, etc

There are legislative requirements in the UK[76] and the USA[72] to assess the risks to the health of the employee and to control exposure. There is practical guidance in HSE and AWS documents.[73,77,78]

The control of exposure to airborne pollutants should follow a hierarchy:

- choosing a process or material that produces less fume/ emissions;
- ventilation of the whole workplace;
- local exhaust ventilation;
- respiratory protection for the individual.

Substitution/Adjustment of Parameters

The first question to consider is whether a method of joining can be chosen that will present a lesser hazard to health. In reality most joining processes are chosen because they are the most effective for the situation and this limits the scope for substitution.

However, it should be noted that of the major welding processes, the production of fume is approximately in the following order: manual metal arc > metal inert gas/metal active gas > tungsten inert gas and oxyacetylene welding > submerged arc welding. (See Chapters 10 and 11 for a description of the welding processes.)

Mechanised welding should be considered, because it can be more easily ventilated by attaching extractor nozzles to the machine, so that fume is removed. The operator is usually some distance away, due in part to the requirement to exclude personnel from the working envelope of the machine.

There is some scope for reducing fume emission within the chosen process. For instance in metal inert/active gas welding, the fume emission rate rises when the arc length is increased. Dip transfer produces less fume than pulsed transfer, whereas spray transfer produces the most. The proportions of toxic substances can vary between the processes. For instance, when welding stainless steel, in the metal-inert gas processes most of the chromium is formed in the less toxic Cr(III) state, whereas in manual metal arc welding it is mostly in the Cr(VI) state. Having chosen the welding process, ventilation will need to be planned.

Factors in Planning Ventilation

To plan what ventilation may be needed in any given situation the following factors must be taken into consideration:

– the nature of the work site;
– the parent metal and the consumables used to weld it;
– the welding process in use.

The work site may be classified as:

– open air, with unrestricted dispersal of fumes;
– general workshop normally requiring general ventilation;
– confined space, in which fumes will rapidly build up.

General Recommendations for Fume Control

General recommendations can be made for the most used processes and materials, see Table 6.1.

Table 6.1. Common gas and arc welding processes – general recommendations[73]

Process	Likely fume emission	Precautions suggested

Gas welding
Manual flame cutting
Tungsten inert gas,
  plasma arc
	Usually below emission
	limits for mild steel
		Work in open air, upwind
		  of the weld if possible.
		Work in workshop with
		  general ventilation.
		Use local extraction for
		  heavy work loads

Manual metal arc welding
Metal inert/metal active
  gas welding
Flux cored arc welding
Flame gouging
Mechanised flame cutting
	Usually greater than
	occupational exposure
	limits
		General ventilation for
		  extremely light work
		  loads.
		Use local exhaust
		  ventilation with good
		  background ventilation.
		Use respiratory protective
		  equipment for work in
		  confined areas

Oxygen arc cutting and
  gouging
	Very high fume levels	Use local extraction with
		  good background
		  ventilation. Check levels

In Table 6.1, a heavy workload is to be judged on numbers of welders, their average duty cycle and the size of each welding operation as related to electrode diameter/current, etc. In the open air, with unrestricted dispersal of fume there is generally no hazard when working on the parent metals specified in the table.

The following is suggested as a general guide to conducting an assessment in order to decide what fume control measures are required:

1 Establish the constituents of the parent metal and consumables, or obtain a written assurance from the manufacturer or supplier about their safety without special precautions.

2 Establish the nature of any coating or contaminant on the surface.

3 For the processes covered in Table 6.1 where the parent metal is mild steel or aluminium, with no toxic material in any coating, use the recommended precautions.

4 Where more toxic materials such as copper, nickel or zinc are involved and work will only be for a short period, take more stringent precautions, such as the use of local ventilation and a dust respirator in the open air or in a general workshop or an air-supplied helmet in a confined space. Estimate the fume

concentration and the dilution required. If work will continue over a period, consult the Welding Institute booklet[79] or obtain expert advice. It may be necessary to have fume levels checked by measurement.

5 For processes other than those listed in Table 6.1 see the Welding Institute booklet[79] or the appropriate chapter in this book.

6 Where highly toxic materials such as cadmium or beryllium will be present during welding, seek information from manufacturers, suppliers or an expert on procedures to ensure safe working and the need for tests to confirm safe conditions before starting work.

7 If any worker shows ill effects which may stem from fume, seek medical advice. Unless fume is ruled out as a cause, check pollutant levels.

8 Arrange regular maintenance and tests of local exhaust ventilation and respiratory protective equipment (see below).

9 Inform workers about the arrangements that are made for their protection and the dangers of the fume, and arrange any necessary training in use of control equipment.

10 Write down a summary of the assessment, recording at least each separate type of situation considered, where the information came from, what action was chosen and the results of any tests.

Fix a date to review the assessment but do so earlier if there is any reason to suspect that it is no longer valid, or if there has been a significant change in the work.

Methods for the Control of Exposure to Fume

Welder position

The natural tendency for a welder is to stand and bend over the work placed on a welding bench. As the hot fume-laden air from the arc rises vertically it enters his or her breathing zone. Thus, if the welder adopts a posture so that his or her head is no longer directly above the arc, the exposure to fume will be reduced. In practice the easiest way is for the welder to work seated, if possible (Fig. 6.1a and b).

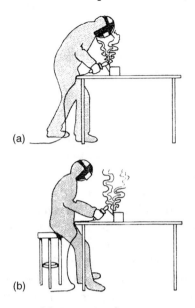

6.1   (a) Excessive fume exposure, (b) improvement in welder position.

General ventilation

Fumes generated by one welder will be distributed around the workplace by convection currents and the fumes released will eventually reach an appreciable concentration unless there is some filtration or removal. The fumes will thus be breathed in by everyone in the shop, although the welders themselves will be exposed to the highest concentrations.

To control this problem, general ventilation is normally installed, extracting air from the general volume of the workshop and not confining airflow artificially to the neighbourhood of the welding work. Enough air must be extracted to reduce fume to an acceptable level in the workshop as a whole.

If background fume measurements are to be made to confirm performance the most adverse conditions should be chosen. This would include the largest number of welders, the highest duty cycle, closing the entrance doors, the longest work period during which fume may accumulate, etc, to give confidence in the ability of the system to cope with all likely loads. Air flow should be well distributed with no stagnant pockets and fresh air must be supplied

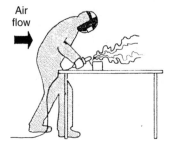

6.2   General ventilation.

(warmed or cooled if necessary) to replace that extracted. Though it may be expensive to install and run, the ventilation system need cause little further interference with the work.

When planning a system, exhaust and inlet air flow should be considered, see Fig. 6.2. Ideally, fumes should be drawn away from every welder's breathing zone, but in practice this will often be difficult to achieve.

Air movement should be between 0.1 and 0.15 m s^{-1}; any greater will lead to complaints about draughts unless the work is physically hard or the weather is very hot. When recirculating air, it should be borne in mind that filtering normally only removes particulates and usually leaves gases and extremely fine particles. The building should have an adequate supply of fresh air. For ordinary work this should not be less than approximately 5 l s^{-1} per person – for welding operations much more will be required.

For short-term use of welding, for example, in repair of machinery, in an area with good general ventilation, the concentration of the fume in the breathing zone of the welder may well be found to be below the exposure limits. In these cases, there is no need for further measures to be taken. However, where general ventilation is not so good, or where a more hazardous material is being welded, although a hazardous background level of fume may not be attained, it may still be desirable to use local extraction as described below to keep the welder's breathing zone clear, or to prevent contamination of the building contents by welding fume.

Local exhaust ventilation

Local extraction is here taken to mean exhaust ventilation where the major part of the airflow is confined to the immediate neighbour-

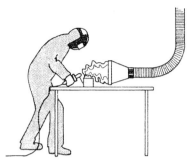

6.3   Local exhaust ventilation.

6.4   Local exhaust ventilation in use (photograph courtesy of Nederman).

hood of the weld. There are two basic types of equipment – the transportable type that is self contained and mounted on wheels, and the fixed fume extraction system installed permanently in the welding shop. In both cases, equipment will normally comprise an extractor nozzle or hood, air hose, fan unit and discharge system; the last either discharges fume-laden air to the outside atmosphere or filters it and recirculates it. Practical advice is available regarding the design of such systems.[77,80,81]

Typical local exhaust ventilation systems are shown in Fig. 6.3–6.5.

Fixed equipment generally has the extractor nozzles or hoods on flexible trunking hanging from the wall, see Fig. 6.5. These systems

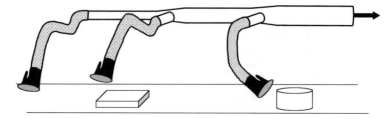

6.5   A fixed local exhaust ventilation system.

require careful planning and installation to ensure that they are balanced so that each work station has adequate air flow.

The success of such a system in extracting fume is largely dependent on the good design of the 'front end', where the design of the nozzle or hood is extremely important.[77] The extractor nozzle should be easily positioned close to the weld; for example some commercial units have magnetic clamps. It should produce an air velocity of $0.5–1.0\,m\,s^{-1}$ over as long a section of the weld as is practicable. Lower velocities will not generally be effective as fume tends to rise at about $1\,m\,s^{-1}$ in convection currents, and higher velocities may remove the gas shield on which the gas shielded processes and MMA welding rely. The equipment should allow the welder a clear view of, and unobstructed access to, the work. The hose should be flexible and resistant to spatter, and the fan unit should be readily portable and reasonably quiet in operation. The duct velocity should be sufficient to prevent deposition of the fume and dust within the system.

Unfortunately, the above requirements tend to be mutually exclusive; a large nozzle extracting large quantities of air will need a large hose and a powerful fan, for instance. The limited coverage obtainable from current designs means that the extractor nozzle must frequently be moved along the work since commercial nozzles are no longer than 0.5 m.

A good design is illustrated in Fig. 6.4. Fume is allowed to rise for a short distance above the weld region, being captured by a relatively large diameter nozzle and hose. A floor standing cabinet, usually fitted with castors, contains a powerful motor and fan and a filtration system to remove fume from the extracted air before recirculating it. A pivoted cantilever arm attached to the cabinet on adjustable mountings enables the collecting nozzle to be positioned simply within a radius of more than a metre. Many modern systems

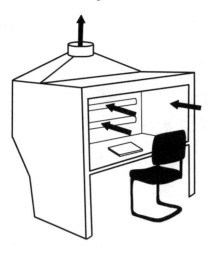

6.6   Extracted bench system.

have an automatic filter cleaning cycle, at the press of a button, so that the fume is dislodged from the filter and is collected in a container beneath the unit for ease of disposal.

Extracted booths and benches

Where the work can be carried out on a bench, this can be permanently fitted with an overhead or rear large extractor hood. The hood may need to be provided with lighting on the inside to illuminate the work area adequately. This form of ventilation can be very satisfactory in suitable cases, but loses its effectiveness if the work is placed so that workers have to lean over it into the fume rising from the weld to reach part of the job.

Face velocities of at least $0.5 \, \mathrm{m \, s^{-1}}$ are required to control the fume and the airflow must direct the fume to the extraction point. A picture of an extracted bench is shown in Fig. 6.6.

On-gun extraction

In the metal arc gas shielded process, or variants using a self shielded wire, the gun may be fitted with an extractor nozzle surrounding the normal contact tip and nozzle assembly, see Fig. 6.7. When correctly designed and operated, effective fume removal is achieved without extra tasks for the operator. However, the bulk

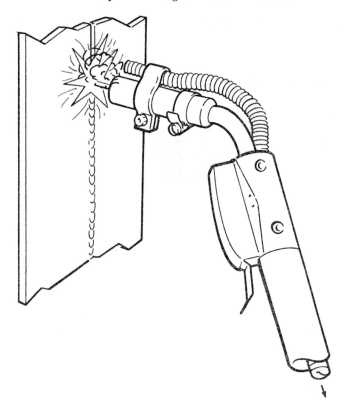

6.7   On-gun fume extraction.

and weight of the gun is increased and the extractor hose may be a substantial encumbrance, so that such a system may be difficult to use except on jobs offering easy access to the joint. The system is basically low volume, high velocity, and it is of no use if too far from the weld. Conversely if it is too close it removes the shielding gas.

Such a device has been marketed both as a complete gun with fume extractor built in and as a clip-on attachment suitable for a range of makes and types of gun. The main application to date seems to have been in connection with flux-cored wire, with which it is able to remove sufficient fume to give entirely acceptable working conditions.

Though very effective in the removal of particulate fume in the arc region, extraction devices do not remove toxic gases, such as ozone and oxides of nitrogen, because they are designed only to filter particulate emissions. Additional general ventilation may be needed to remove these gases.

Treatment of Extracted Air

Direct discharge

Extracted polluted air may be discharged to the atmosphere outside the workshop. Though simple in principle there are three potential problems. First, especially with local ventilation, an extensive hose or trunking system may be needed in a large building. Second, especially with general ventilation, the make-up air drawn in to replace that extracted may need to be heated to raise it to a reasonable temperature in the winter, or to be cooled in the summer. Third, it is possible that it would be unacceptable to pollute the outside atmosphere, depending on the emissions and the siting of the factory. All these difficulties can be overcome if it is possible to clean the air by removing pollutants and then to recirculate it.

Filtration

A filter will effectively remove particulate matter, but will not deal with asphyxiant or pollutant gases, oxygen enrichment or explosive hazards. Where a filter is used, regular cleaning and/or replacement will be needed: the person who carries out this operation may need protection from toxic dust and used filters must be safely disposed of. Some fume extractors have a cleaning cycle, where the unit will dislodge the fume from the filter and dump it into a container for disposal. The administration and financial implications of the system should be taken into account when it is selected.

Electrostatic precipitator

In an electrostatic precipitator air is passed between two flat metal plates, typically spaced by 10 mm, connected to a high voltage (about 15 kV) low current supply. Any particle passing between the plates becomes electrically charged and is then attracted to one plate or the other by electrostatic forces. The plates are long enough in the direction of air flow to intercept all relevant particles before the air stream can carry them clear of the plates. In practice a series of plates, connected alternately to positive and negative poles of the high voltage supply, increases the capacity. When the plates have collected so much dust that a discharge can occur between them, the filter plates are removed from the unit and brushed or washed down. The power supply has a very limited current and is relatively safe to touch, but nevertheless interlocks are provided.

Electrostatic precipitators may be applied to recirculate air from either local extraction or general ventilation systems. They can be 92–98% efficient for particles of order 5 μm, but some materials do escape.

Fresh air supply

In combination with extractors, particularly in confined spaces, clean filtered fresh air may be fed via suitable trunking to the general area of work. The fans for this duty are similar to those for extraction but usually of somewhat greater capacity. An extractor draws in air from all directions, but the air supply is more directional and should be arranged to blow towards the welder and carry fumes to any extraction vents.

Respiratory Protective Equipment

In specialised cases, respiratory protective equipment, RPE, will be required. These cases include welding in an area with a high concentration of fume, or where the fume is highly toxic. In places where there is an immediate danger to life from inhaling the air, the RPE must provide air of breathable quality from an independent source. It is very important to ensure that any RPE is compatible with the other protective equipment required for the work being undertaken (e.g. welders helmet, hearing protection). A typical example is shown in Fig. 6.8.

When purchasing RPE, many items will have a specified protection factor (PF). This is the factor by which the device will reduce the contaminant, and hence a PF of 10 will protect the wearer in an atmosphere that contains a contaminant at 10 times its exposure limit. However, it must be stressed that training is required to ensure that the protection is used correctly.

Table 6.2 summarises the various types of respiratory protective equipment and their typical uses. Further advice may be gained from the literature.[84,85]

6.8   Respiratory protective equipment (photograph courtesy of Nederman).

In all cases, the protection offered is dependent on the attainment of a good fit. This is achieved by the selection of the right equipment and training of the personnel. Note that the RPE that requires air to be supplied to the welder requires air of an acceptable quality, to BS 4275[82] or to CGA 7-1 grade D.[83]

It is important to maintain an adequate stock of spare or replacement parts. RPE will, in most cases, require a maintenance programme, which will consist of cleaning, disinfection and performance tests.

Quantifying the Fume Extraction

Measurement of fume

The techniques of welding fume measurements are now well established and published as standards.[86–92] Careful adherence to the prescribed procedure will minimise variations due to experimental method in the results obtained, leaving an inevitable spread from the vagaries of formation and dispersal of welding fume.

Background measurements indicate the amount of fume present in the air throughout the workshop, and therefore that to which all workers would be exposed. These should be substantially less than any exposure limit if possible, to allow for extra fume near welding operations.

Table 6.2. Respiratory protective equipment

RPE	Protection achievable	Used for
Disposable filtering face piece respirator	Potentially 20 times the exposure limit	Mainly used for protection from dusts
Half mask respirator with filter	Potentially 20 times the exposure limit	Provides no protection for the eyes
Full face mask respirator and filter	Potentially 40 times the exposure limit	Greater protection than the half mask, and gives protection to the eyes
Powered respirator with helmet or hood	Potentially 40 times the exposure limit	Greater protection for the head. Filtered air can cool the face and assists with breathing
Powered-assisted respirator with full facemask	Potentially 40 times the exposure limit	Can be purchased customised for welding
Air-supplied mask, by breathing or by fan assistance	Up to 40 times exposure limit	Limited to about 9 m by the hose
Compressed air line mask and hood	Can achieve protection up to 2000 times the exposure limit	Air must be to acceptable quality BS 4275, CGA 7-1 grade D[82,83]
Self-contained breathing apparatus	Up to 2000 times the exposure limit	For the most hazardous situations, especially where the workplace atmosphere does not support life. Requires high standard of maintenance and training

Breathing zone (BZ) measurements are chosen for the majority, to indicate the welder's exposure. Because a helmet affects air flow around the welders head, the sampling device is attached inside the helmet to obtain a true sample of the air breathed.

Particulate fume is measured by collecting it on a previously weighed filter through which air is drawn at a known rate by a small battery-powered portable pump. After the welder has worn it for a timed period of about an hour, the filter is removed for laboratory measurement. The filter is weighed on a sensitive balance; from the increase in weight above that of the clean filter, the total accumulated fume is measured. Knowing the air flow rate and the time, the fume concentration can be calculated in milligrams per cubic metre

($mg\,m^{-3}$). This will be a measurement of total inhalable dust, which is what is normally done under welding conditions.

It is theoretically feasible to measure the respirable dust separately from the total inhalable dust by using a preliminary filter to capture the larger particles above about 7 μm diameter. These would then not be weighed, allowing the weight measure to be that of respirable dust which is small enough to go down into the lungs.

Chemical analysis of the deposit on the filter enables the concentration of individual elements and compounds to be found. Gases may be measured by a chemical reaction in a one-shot disposable gas detector tube or by a special purpose analytical instrument. The amount present is expressed as a concentration in parts per million (ppm).

Since skilled personnel and special equipment, backed up by appropriate laboratory facilities, are needed to make meaningful fume measurements, many employers choose to use an outside organisation providing this service. Whoever does it should be competent, although no precise qualifications are laid down by the regulations. Evidence of previous experience should be requested, and if there is any doubt, checked with the enforcing authorities. Alternatively, some laboratories operate under an approved quality control system.

Calculating the Reduction in Fume Level Required

Having a measurement of the total fume generated may be sufficient in order to calculate the fume control required, even when one of the constituents has a low exposure limit.

An example

A welding consumable produces a fume composition that typically contains 14% chromium as Cr(III).

The exposure limit for Cr(III) is 0.5 $mg\,m^{-3}$.

Therefore the welding fume will reach this exposure limit when the total fume is (100 × 0.5)/14 = 3.6 $mg\,m^{-3}$ (to two significant figures).

Thus if the total fume is less than 3.6 $mg\,m^{-3}$, the concentration of Cr(III) will be below the exposure limit.

Manufacturers of welding consumables will frequently give this information, in terms of a figure to which the total fume should be controlled in order to keep the more toxic constituents within the limits.

Maintenance of Equipment

All equipment for the prevention of exposure to substances hazardous to health must be maintained. On a weekly basis equipment should be checked visually for signs of damage, wear or malfunction. Components that are found to be faulty, worn or damaged should be replaced promptly. For instance, local exhaust ventilation (LEV) that has a flange as part of its hood design can suffer a loss of efficiency of around 50% if the flange is lost.

At intervals, local exhaust ventilation must be inspected and have its performance checked by measurement. This would include checking that the airflow velocity and volume was still within specification, an electrical check, and a check that the nozzle or hood is in good order. The pipework should be inspected and checked for leaks, accumulations of dust, etc. This is particularly important in the elements of the system that are flexible – moving pipes around can easily lead to damage.

RPE must have its own regime of maintenance, which will include regular visual inspection of such items as the hoses, seals, filters (if any), and more formal checks of such items as fan performance (where applicable).

Training the Welders

No programme for control of exposure to fume is complete without giving the welders the information, instruction and training that they need. They should be taught:

– Hazards to health
– How to modify or operate the process to minimise the risk
– How the protective devices work and the best way to use them
– How the general ventilation system operates to assist the control of exposure
– Limitations of the control measures
– How to recognise signs of failure
– What to do in the event of failure of the control measures
– How to keep the equipment in efficient working condition.

7

Radiation

Radiation includes light, heat and ionising radiation, all of which are found within, or associated with, the welding environment.

Non-ionising Radiation

Ultraviolet, visible and infrared radiation all belong to the electromagnetic spectrum, part of which is shown in Table 7.1.

Sources of Radiation

Arc welding produces large quantities of light, from the ultraviolet (UV) to infrared radiation. UV radiation from a welding arc is intense – for example a metal inert gas (MIG) weld using helium gas running at 300 A typically produces 5 W m^{-2} in the UVB and UVC at a distance of one metre. This is many times the intensity of the sun at noon. The visible radiation is also intense. Oxyacetylene welding produces less UV light, but still produces substantial quantities of visible and infrared radiation.

The lasers used in welding are typically carbon dioxide, at a wavelength of 10.6 μm, Nd/YAG at 1.06 μm and excimer lasers which operate in the UV region. The radiation from lasers is intense and the beams have very small divergence, so that they travel long distances with very little reduction in power density.

Health Effects

The UV radiation from a welding arc can cause severe burning to any unprotected skin. Exposure of the eyes to this radiation causes a condition known as 'arc eye' or 'welder's flash' which is an inflammation of the cornea. This condition typically appears some hours

Table 7.1. Part of the electromagnetic spectrum

Approximate wavelength	Type of radiation	Common name
<100 nm	Ionising radiation	X-rays, gamma rays
100–280 nm	Ultraviolet light	UVC
280–315 nm	Ultraviolet light	UVB
315–390 nm	Ultraviolet light	UVA
390–760 nm	Visible light	
760 nm–1.4 µm	Infrared radiation	IRA
1.4–3 µm	Infrared radiation	IRB
over 3 µm	Infrared radiation	IRC

after exposure, when the eyes become red, watering and painful, often with a gritty feeling. The eyes may become sensitised to light. A person who believes that he or she has arc eye should seek medical advice, to ensure that there is no foreign body in the eye, but an uncomplicated arc eye condition will heal spontaneously.

UV light can burn unprotected skin causing blistering; these burns should be treated as normal thermal burns. Long-term exposure to UV radiation can cause premature ageing of the skin, cancers and, for the lens of the eye, premature yellowing and cataract.

Intense visible light can injure the eye. Long-term exposure can cause degeneration of the retina. Infrared radiation can cause skin burns and cataracts in the long term.

At one time it was customary to paint welding booths a dark colour, but it is now recognised that the colour is not important, provided it is a matt finish that will not reflect the radiation (particularly the UV). There are considerable advantages to choosing a light colour, since it helps to improve the general illumination of the area. Paints with titanium dioxide and zinc oxide have low reflectance of the UV.[93]

Protection from UV–Visible–IR Radiation

The welder is generally protected by clothing and those further away should be protected by screens and curtains. Clothing should cover all the skin, to protect not only from radiation, but also spatter, electrical contact and sparks. Depending on the demands of the task this clothing will include a boiler suit and cap, gloves, shoes or boots. For heavy work aprons, shoulder covers, jacket and leggings or spats may be added.

Personal clothing should be made from wool or cotton, so that it cannot melt. Additional protective clothing, such as aprons, are fre-

7.1   Helmet with electronically controlled filter, powered by
solar cells above the viewing aperture, with knob adjustment of
the effective shade number (photograph courtesy of Racal Safety).

quently made from leather. Protective clothing should not have
external pockets, unless they have a closeable flap to completely
overlap the opening. Clothing should be designed to avoid openings
or folds where spatter can lodge. Specific safety requirements are
given in the Standard BS EN 470–1.[94]

Gloves should protect against electrical contact and radiation.
Most welding gloves are made from leather. Cotton gloves may be
worn inside as liners. The seams should be inside (to avoid the
thread being burnt) and the materials should be suitable for the
temperature range to which they are exposed. A standard is currently in draft.[95]

Protective screens should be used to shield adjacent welding stations from one another, and also others in the environment. Screens
are available in a wide range of colours, each being suited to particular welding operations. The materials from which screens are
made should be non-combustible or flame resistant.[96,97]

Eye and face protection must comply with the relevant standards.[98,99] In the USA eye and face protection should comply with
Z87.1.[100] Shade numbers should be chosen from the shade selector.[101] Welding helmets are available that automatically darken and
are manufactured to BS EN 379, see Fig. 7.1.[102]

Ionising Radiation

Ionising radiation is produced from two main sources:

- radiation generators (e.g. X-ray and electron beam equipment) and
- radioactive substances

both of which are likely to be encountered in association with the testing of welded joints.

Health Effects of Ionising Radiation

Acute effects appear when the radiation is above a threshold. The severity of the effect depends on the radiation dose. Acute effects include radiation burns and cataracts. It should be stressed that these effects are only found at high radiation levels and so are normally only seen after a serious incident.

The longer term effects are the result of smaller doses. For these effects there appears to be no threshold dose. There is now a considerable body of evidence that radiation can induce various forms of cancer. So far no genetic effects have been demonstrated conclusively.

Quantifying Ionising Radiation

In the UK the dose is measured in sieverts (Sv) the energy absorbed per unit mass multiplied by a weighting factor that adjusts the dose to take into account the degree of damage that each type of radiation can do. The alternative units are rem, 'roentgen equivalent man', where 100 rem = 1 Sv. In both the UK and the USA there are dose limits for employees that are related to age, parts of the body affected and previous dose history.[103,104]

Radiation Generators

Radiation generators use an electrical source at a high voltage. Electrons fall on a target and generate X-radiation. X-rays are also part of the electromagnetic spectrum, lying beyond UV (see Table 7.1). X-rays are routinely used for radiography to reveal imperfections inside welded joints.

It is desirable for X-ray generators to be housed in purpose-built rooms with doors that are interlocked to the power supply. In this way it can be made impossible for a person to enter the enclosure while the radiation is being emitted. Procedural controls should require a search before closing the room. Audible and visible warnings are used to signal impending X-ray release and panic buttons are installed in the room, to allow anyone accidentally left inside to halt the operation. The room is lined with absorbent material such as lead or concrete, to ensure that there is no discernable radiation outside the enclosure.

Radioactive Substances

Radioactive substances are atoms that decay naturally. They can give off alpha particles, beta particles and gamma radiation. Unlike X-ray sources they cannot be turned off, so their control is more difficult. Sources for industrial radiography such as iridium 192 are emitters of gamma radiation and they can be used to radiograph thick sections of steel and other metals. These, too, are used inside shielded enclosures, but since the sources cannot be turned off electrically, they are housed in shielded containers. From the container, the source is projected through a guide tube to the point of use, then retracted. Procedural controls will be needed so that checks are done each time the source has been used to see that it has indeed returned to its container. Serious accidents have occurred when sources have been accidentally left inside the guide tubes.

A common source of alpha particles in the welding environment is thoriated tungsten, used in many TIG welding electrodes. Alpha particles do not travel very far, and are effectively stopped by the layer of dead skin on the outside of our bodies. However, alpha particles have a great deal of potential to cause harm. If they are released inside the body, as a result of inhaling or ingesting a dust that emits alpha particles they can cause damage to the lungs or digestive system. Measures must be taken to minimise the risk of this happening. This requires control of dust when the electrodes are ground.

Setting up a Facility

An employer intending to set up a radiographic facility should seek professional help and liaise with the enforcement authorities. Authorisation will be required and the facilities will need to be constructed and operated to the required standard.

8

Noise and Vibration

Noise

For the purposes of industrial health and safety, noise includes all forms of sound that can be perceived. Sound is generated when objects vibrate and these vibrations give rise to small changes in the pressure in the air adjacent to the vibrating object. The resulting pressure waves travel away from the vibrating object. The frequency with which the vibrations take place will determine the frequency of the pressure waves; high frequencies are perceived as high pitched sounds, whereas low frequencies are perceived as low pitched. The perceived loudness of the sound is related to the difference in pressure between the peaks and troughs of the pressure waves. Particularly noisy processes in the welding environment are plasma arc cutting and gouging, arc cutting and gouging, chipping, grinding and engines and generators.

Hearing Mechanism

The human being has a complex sensory mechanism for receiving and analysing sound. It starts at the outer ear, which collects these pressure changes and funnels them down the ear canal to the ear drum. The ear drum vibrates in response to the pressure changes and the vibrations are transmitted via some tiny bones to the cochlea, which is a structure that is filled with fluid and is lined with thousands of short hairs embedded in its wall. When these hairs vibrate in response to the sound, signals are sent to the brain. A schematic diagram is shown in Fig. 8.1.

Noise has some immediate effects. It causes difficulty in communication, increases stress levels and thus increases the likelihood of accidents. Exposure to loud noise can cause temporary dullness

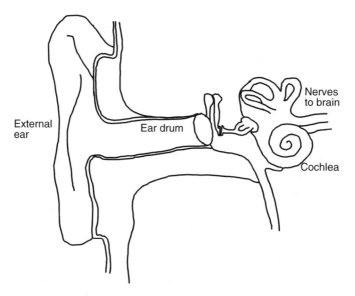

8.1   Schematic of the hearing mechanism.

of hearing and a rise in the threshold of hearing (known as a 'threshold shift'). In favourable cases this loss is temporary and the hearing recovers gradually over a period of hours. However, if exposure to noise is repeated, or if the exposure is to an excessively loud noise, then the damage may be permanent. This is because the hairs in the cochlea can be physically damaged by repeated or excessive exposure to noise. The loss of hearing that results cannot be rectified by a hearing aid. Excessive exposure to noise may also cause tinnitus, which is a distressing condition.

The probability that long term damage to hearing will occur in an individual depends on three main factors:

– noise level and frequency characteristics of the sound
– duration of exposure
– susceptibility of the individual.

The loudness of sound is measured in decibels (dB). The human ear is not equally sensitive to all frequencies or pitches of sound. Hence measurements are generally filtered so that the measuring instrument has a response equivalent to the human ear. When this has been done, it is said to be 'A' weighted, and the unit is then expressed as dB(A) in the UK or dBA in the USA.

Table 8.1. Equivalent noise exposures

Time (hours)	Exposure level (UK)/(dB(A))	Exposure level (USA)/(dBA)
8	90	90
4	93	95
2	96	100

The decibel measure is logarithmic, so that small changes in the decibel level can represent large changes in the noise level. A rise of 3 dB represents a doubling of the noise and so exposure for 8 hours at 90 dB(A) would be equivalent to 4 hours at 93 dB(A). (Note: in the USA, with the exception of the armed services, a 5 dB exchange rate is used so that 8 hours at 90 dBA is equivalent to 4 hours at 95 dBA. There are proposals to change this.)

In order to quantify the risk of noise-induced hearing loss, noise exposure is measured as a time-averaged exposure. A person who is exposed to 90 dBA for 8 hours, or its equivalent, is said to have had a daily personal noise exposure of 90 dBA. Some trade-off can be made between noise level and time, so that the exposures shown in Table 8.1 are approximately equivalent. This trade-off cannot be taken to extremes, since very high noise levels can do permanent damage in a single exposure.

Legislative Requirements (UK)

There is a general requirement that employers shall reduce the risk of damage to the hearing of their employees to the lowest level reasonably practicable.[105,106] If the daily noise exposure of any employee is likely to reach or exceed 85 dB(A) then the employer shall carry out a noise assessment to identify the employees exposed, so that they can take the measures required by law. The employer shall inform the employees of the possible risk to hearing, what steps they should take to minimise the risk, and how they can obtain personal ear protectors. Ear protectors should be available on request. The employer should provide adequate storage for them, and ensure that they stay in efficient working order. Training will be needed to ensure that the employees know how to use the protectors effectively.

If the daily noise exposure of any employee is likely to reach or exceed 90 dB(A), or he or she may be exposed to a peak sound pressure of 200 Pa or above, in addition to the measures in the previous paragraph, the employer should reduce noise exposure of the employee as far as is reasonably practicable by means other than by personal hearing protection. This has the effect of requiring an employer to consider reduction of noise at source as the first priority.

The employer should delineate areas where the risk of exposure remains at a daily noise exposure of 90 dB(A) or above, or the peak sound pressure level or above, as ear protection zones. These zones should be marked with signs indicating that hearing protection must be worn. The wearing of hearing protection in these zones is mandatory, and the employer must ensure as far as is reasonably practicable that his or her employees comply. The hearing protection should be chosen such that when properly worn it reduces the risk of damage to hearing below the daily noise exposure of 90 dB(A) or peak sound pressure level of 200 Pa.

The employer should record the noise assessment and review it if there is any reason to believe that it is no longer valid, or there is a significant change in the work to which it relates.[105] Most employers are likely to need some specialist help in carrying out a noise survey. Competent persons are able to conduct noise surveys efficiently and to advise on noise control and hearing protection measures that may be employed.

Audiometry, the assessment of the hearing of individuals, is not a legal requirement. However, where noise has been assessed to be a significant risk, an employer should give serious consideration to testing the hearing of his or her employees. This could be done:

- on recruitment – to give a base-line
- at intervals during employment, to check that no deterioration has taken place
- on leaving the organisation.

Legislative Requirements (USA)

The permissible exposure limit in the USA is 90 dBA averaged over an 8 hour period.[107,108] Regulations require a continuing and effective hearing conservation programme for all employees whose exposure to noise is an 8 hour time-weighted average of 85 dBA or more. All noise above a threshold of 80 dBA must be included in the assessment. The employer must institute feasible engineering controls if exposures remain at >90 dBA for an 8 hour time-weighted average. They must also institute a hearing conservation programme which including monitoring exposure, audiometric testing, audio-gram evaluation, hearing protection for employees with a standard threshold shift, training, education and record keeping. (Note: the construction industry standard 29 CFR 1926.52[108] does not contain a requirement for a hearing conservation programme.) Hearing protection must attenuate noise to the levels defined in the regulations. For any employee who already has a threshold shift, hearing protection is mandatory. Notices 'Noise can damage health' should be placed on equipment to which it applies.

Hearing Protection

Where noise cannot be reduced to acceptably low levels, hearing protection is required. In order to choose a type that is adequate, the noise will need to be analysed to characterise its frequency spectrum to quantify noise levels at different pitches or frequencies. The literature from the manufacturer of the hearing protection can then be consulted to choose protection that is adequate. It is good practice to choose hearing protection that is in excess of what is required, to take account of the possibility that employees are likely to fit it incorrectly at times and may forget to wear it for part of the day. Table 8.2 illustrates the main types that are available and the advantages and disadvantages of each type.

When purchasing hearing protection, choose those with seal material that is capable of withstanding heat and spatter impact. Information should be available from the manufacturers, who should be manufacturing their goods to a recognised standard.[109,110] The hearing protection needs to be compatible with other personal protective equipment needed by the welder, Fig. 8.2.

Vibration

Vibration can cause a range of health problems depending on the part of the body exposed. For the welder, the most likely exposure is to hand–arm vibration from the use of percussive tools such as chipping hammers or needle guns, and rotary tools such as grinders.

Studies have suggested that the hand–arm system responds differently at different frequencies of vibration and to take this into account a weighting system is used. The frequency range of vibration that appears to be relevant is 2–1500 Hz, with the most important appearing to be 5–20 Hz.

Table 8.2. Hearing protectors

Type of hearing protection	Advantages and disadvantages
Ear muffs	Can provide a high level of attenuation and physically protects the ear. Easier to achieve the required noise reduction than with other types of protection. Can easily be seen. It can be difficult to get a good fit with other personal protective equipment such as glasses or eye protection, and safety head wear
Foam ear plugs	Only effective if fitted well. Comfortable for long periods of use. Limited life. Workers need clean hands to roll them prior to insertion. Easy to carry and store. Compatible with glasses, etc
Premoulded ear plugs	An air-tight seal is required for good performance and this is not always possible. Easy to carry and store. Compatible with other personal protective equipment. Easier to use than foam plugs and easy to keep clean
Canal caps or semi-aural devices	These provide less protection than the above methods Easy to use, good for intermittent use. Compatible with most other personal protective equipment

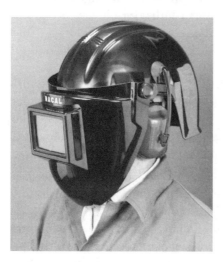

8.2   Ear muffs attached to a combined welding and safety helmet (photograph courtesy of Racal Safety).

In a similar manner to noise, an equivalent value of vibration over time is measured. This value is reported as an A(8), which represents an 8 hour frequency-weighted root mean square acceleration entering the hand–arm system. The unit of measurement of A(8) is metres per second squared ($m\,s^{-2}$).

Initial symptoms of vibration damage are often set off when the person is cold or wet, and can be as little as whiteness of the fingertips (hence the term 'vibration white finger'). If exposure continues the affected area grows larger and there may be numbness or pins and needles. On recovery, the area may become red and painful. Prolonged exposure can lead to damage to the nerves in the hands and to the muscles, bones and joints of the arm. Once these structures have been damaged, the effects are generally permanent.

The following factors are important in determining whether vibration is likely to cause harm:

- magnitude and frequency of the vibration
- daily exposure and the pattern of exposure and breaks
- cumulative exposure
- grip, or force applied to the work tool or workpiece, and the method of work
- user's posture
- area and part of the hand in contact with the vibrating system
- type or hardness of the workpiece or tool
- individual's susceptibility, including factors that affect circulation, such as smoking and medication
- climate.

Legislative Requirements

Neither the UK nor the USA currently have specific legislation relating to vibration (although in the USA, a Regulation 29 CFR 1910.900 is in draft[111]). However, in both countries there is a general duty of care to provide a work place and work activities that do not cause serious physical harm.[1,5] This general duty of care is used by the enforcing authorities to ensure that the risk of harm from vibration is addressed.

In the UK, the Health and Safety Executive recommend that where workers are exposed to A(8) levels exceeding $2.8\,\mathrm{m\,s^{-2}}$ preventive measures are taken, and health surveillance is introduced.[112]

Mitigating Methods for Vibration

One of the primary methods of reducing vibration damage is to replace high vibration tools by tools that produce less vibration. Manufacturers of vibrating equipment are required to give figures for the measured vibration levels, and these are done under strictly controlled conditions in accordance with a recognised standard.[113–115] It is important to realise that these measurements are often significantly different from the vibration levels actually experienced when the tools are used in the field. Also, take into account that an efficient tool will need to be used for a shorter time than an inefficient tool – thus any vibration reduction strategy that compromises efficiency could result in higher exposures.

The differences between good tools and poor tools can be quite dramatic. For instance, in one study it was shown that the time taken to be exposed up to the action level, $2.8\,\mathrm{m\,s^{-2}}$ A(8) for a chipping hammer was only 5 minutes in a poor tool, and 4 hours in a good tool.

It is suggested that, if an employer has vibrating or percussive tools in the high risk category the following measures are adopted:

- Remove the hazard by automation or change of technology.
- Institute a strategy to ensure that lower vibration tools are purchased.
- Minimise the vibration hazard to the hands.
- Reduce the exposure time.
- Maintain the equipment to reduce the vibration level.
- Give instruction on correct operating techniques.

Workers should be educated to recognise the warning signs. They should be taught that they may be at risk if they get tingling or numbness during or after using a vibrating tool or machine and that they should report these symptoms promptly. As a rule of thumb, any piece of equipment that causes tingling or numbness after 5 to 10 minutes of continuous use should be regarded as suspect. The employees should be told about measures that they themselves can take to reduce the risk – keeping warm, not smoking and taking exercise.

The employees should use the right tools, use no more force than is necessary, take breaks and keep machines in good working order. There is no effective personal protective equipment for this hazard. Gloves that claim to deaden vibration are likely to be ineffective in the frequency ranges that are most dangerous and in addition the user may be forced to grip the tool more tightly, actually making the problem worse.

9

Mechanical Hazards

The mechanical hazards presented in welding and cutting are common to most engineering work, but there is a change of emphasis: particular attention should be given to the following points.

Safe Platforms

When working where a fall to a lower level is possible, a safe working platform should be provided. Open edges should be protected by handrails and toeboards; where appropriate, a safety belt should be worn. While fall protection is usually a routine matter, it must be recognised that electric shock can easily cause falls, so that arc welders are at greater risk.

Obstructions

Working areas should be kept free from obstructions as far as possible. This is particularly important where a welder or cutter may have to move while working, since eye protection filters restrict vision. The absence of accumulations of rubbish, slag, etc, makes it easier to avoid damage to hoses and cable and to see any damage which does occur.

Mechanical Lifting

A number of unsafe situations can arise during lifting of work. Wire ropes may be damaged by hot work or sharp edges or even by welding current passing through them (see Chapter 11). Work may have been built up to a weight in excess of the safe working load of the lifting gear or work positioner, or its centre of gravity may be in an unsafe position. Tack welds or untested welds, such as those holding temporary lifting lugs, may part when they bear stress on lifting.

Lifting operations should be planned so that they are safe.[116] This will include a person selecting the correct lifting equipment (e.g. the correct sling) to match the load, and performing the lift safely. Only suitably trained persons should be permitted to use cranes and other lifting aids.

Manipulators and Positioners

Rotary tables or rollers used to position work so that all welds can be made in the best position, for example flat or horizontal–vertical, present a number of hazards which have been overlooked in the past by welders. A safe working load, maximum work dimensions and allowable out of balance load should be established to avoid over-stressing the equipment. Where work is moved during welding, the return path for welding current must be considered; if a cable is used, will it coil up safely as work proceeds or will an assistant be needed to guide it? If a cable is not used the return current must not be allowed to pass via the bearings and damage them, but through a proper slip ring and brush, usually provided on equipment intended for this duty. Manipulators should be securely fastened to a sound foundation or a large unbalanced load may cause them to tip over suddenly during welding.

As with any motor-driven equipment, the welder should have ready access to an emergency stop button.

Where circular work such as drums or pipes is rotated on rolls, suitable precautions should be taken as required to detect and rectify any tendency for the work to creep along its axis of rotation. Work with holes in the outer diameter, or projections such as stub pipes, may foul rollers or fixed plant during rotation.

Particular care is needed to avoid starting with a safe piece of work, which is then built up to a weight or size exceeding the capabilities of the equipment in use.

Wire Feed Units

Wire feed units, used particularly in gas-shielded metal arc welding and in mechanised welding and surfacing by a range of processes are often capable of exerting enough force to drive the sharp end of the wire into the operators hand. Operators should not place their hand over the gun and pull the trigger to check the gas flow: they should use the gas purge facility and/or place their hand clear of the wire.

Grinding

Portable grinding tools must be adequately maintained for safe operation. Electric tools should be checked for earth lead continuity or double insulated as appropriate (see Chapter 1).[117,22] Air tools must be used only from an appropriate airline with a sound hose (never from an oxygen supply). Wheels must be correctly chosen to suit the speed of the tool and correctly mounted; this is a legal requirement in the UK[118] and the USA.[119]

Robots

The applications of industrial robots in welding are steadily being extended, mainly into areas where quantities are sufficient to justify an improvement over manual operation but not great enough to warrant a special purpose machine. Resistance welding and gas-shielded metal arc welding are the two processes most often used in conjunction with robots. In both, speed of operation, especially in transit from one weld to the next, plays an important role in the economics. This potential for high velocity and acceleration in movement brings the risk of injury to anyone who gets in the way. Guidance is available for robot installation and operation.[120-122]

During normal operation, therefore, no-one should be within the volume within which the robot can traverse. This can be achieved by surrounding the robot with a mesh perimeter fence, with panels providing access to the area interlocked. The interlocking must be designed to hold the machine stationary until all the panels are in place; movement must not then start without initiation by a further separate action. Opening a gate to the robot should cause the motion of the machine to cease, and not to be resumed until the system is reset (Fig. 9.1). If the cage is not large enough to encompass the entire working envelope of the robot, then hardware stops should be installed to prevent the robot from breaking through the cage. Software stops are not sufficient.

The guards also provide reasonable protection against workpieces being thrown out and should prevent any arc being directly visible to anyone in the neighbourhood. If it would be possible for someone to be inside the fence when the access door or doors are closed, pres-

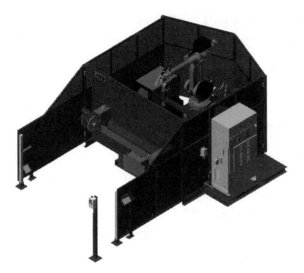

9.1   Robot Flexarc 250R, welding cell based on robot IRB 1400 and positioner 250R, with interlocked gate (photograph courtesy ABB).

sure mat switches or light beams can detect their presence and lock out the power.

Loading and unloading work is commonly achieved without the need to enter the enclosure by adopting a turntable carrying two or more jigs (Fig. 9.2 and 9.3). Fume extraction is often needed because of the high duty cycle which can be achieved. If an overhead extractor hood is not suitable, it is possible to fit a local extractor on the robot, often leaving the last one or two joints unencumbered. When this has been done, robot technology offers a considerable improvement in safety and working conditions over manual operation; no physical effort is needed to position equipment and no worker is directly exposed to the heat, fumes and light which may be emitted from the weld zone. Training is required to operate the equipment satisfactorily and safely.

On the occasions when a worker must be in close proximity to the robot, for such jobs as cleaning, care must be taken that the robot has been immobilised and cannot move until the job has been completed. This safe system of work might for instance be implemented by such means as a captive key system.

All robots should be fitted with some form of emergency stop; all those authorised to initiate operation must be instructed as to how to resume normal operation, for example how to restart at the same point or how to reset to the start of the programmed cycle. As a

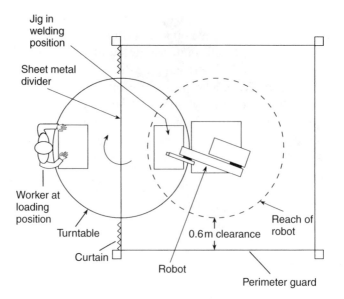

Jig in welding position

Sheet metal divider

Worker at loading position

Turntable

Curtain

Robot

0.6 m clearance

Reach of robot

Perimeter guard

9.2   Plan of robot welding enclosure.

9.3   Robot welding cell, of similar layout to Fig. 9.2, but with solid walls and a tinted glass turntable divider (photograph courtesy Torsteknik).

number of accidents have occurred when components have been seen to be misplaced and the robot continued its cycle as soon as the fault was rectified, all concerned should receive adequate instruction on a procedure for making the equipment safe before entering the hazard zone. The procedure must take account of the need to ensure safe working whether the emergency stop had been activated or not, and whatever stage in the operating cycle has been reached before interruption. If there is any doubt, damage to work in progress or to equipment must be accepted as preferable to even a slight risk of personal injury.

For setting up a new job, or for fault diagnosis, it will often be necessary for a worker to be close to the weld point with the robot operational. For instance, the robot may be programmed by the operator steering the gun in the desired path around the work via a manual control box in the teaching mode, the equipment recording the required motions in a memory for subsequent replay. Where possible there should be some 0.6 m clearance between the maximum extension of the robot and the guard mentioned above to prevent the operator being crushed between robot and guard. Pressure mats on the robot arms can provide extra safeguard, but it is difficult to cover all potentially hazardous situations.

Emergency stop buttons should be fitted so that one at least is in reach from any point in the hazard zone. Control consoles should preferably be placed in such a position that the operator is not within the hazard zone, or if this cannot be avoided, so that the operator cannot be crushed between the robot and the console. The teaching or other manual operation mode should be restricted to a slow speed, say $0.25 \, \mathrm{m \, s^{-1}}$ and the control box should have a dead man's button which has to be held down before any movement can start and continue.

A further order of complexity arises where welding robots are integrated into complete production lines, as in computer integrated manufacturing CIM systems. Careful consideration will be needed to ensure safe access to the robot cell for setting up or repair without incurring total disruption of the entire installation.

Eye Protection

Workers carrying out deslagging, chipping or grinding should use appropriate eye protection.[99,100] The British Standard provides for several grades of protection; these are summarised in Table 9.1.

Table 9.1. Eye protection[99]

Grade to resist	Marking
Liquid droplets	3
Liquid splashes	3
Dusts, >5 μm	4
Gases and fine dusts, <5 μm	5
Molten metals and penetration of hot solids	9
Increased robustness	S, tested by 22 mm ball at 5.1 m s^{-1}
Low energy impact	F, tested by 6 mm ball at 45 m s^{-1}
Medium energy impact	B, tested by 6 mm ball at 120 m s^{-1}
High energy impact	A, tested by 6 mm ball at 190 m s^{-1}

Protective Clothing

In most welding shops, safety footwear will provide valuable extra protection. Where processes involving substantial amounts of molten metal, such as electroslag or thermit welding, are in use, a pattern designed for use under foundry conditions will be more appropriate. On many sites safety helmets are required; some patterns are available which combine or can be worn with a welding helmet.

Part 2

Processes

10

Gas Welding, Cutting and Preheating

Processes that are included in this chapter are: gas welding; oxy-acetylene welding; gas cutting; oxygen cutting; oxygas cutting; flame cutting; gouging and lancing; and powder cutting.

The Gas Flame

A fuel gas and oxygen or air are fed into a blowpipe where they are mixed, and the mixture is delivered through a nozzle where it burns. The flame produced by this combustion consists of two zones, an inner cone-shape, and an outer zone in which combustion products act to shield the weld area from the atmosphere, see Fig. 10.1.

For welding, the fuel gas is acetylene, with oxygen to achieve a high enough flame temperature to melt metal; for cutting, described below, and for general heating, the fuel gas is often propane. Brazing and soldering do not require such high flame temperatures and other fuel gases may be used, sometimes with air instead of oxygen; as air is only about 21% oxygen, with the rest inert nitrogen and argon, flame temperatures are much reduced compared with oxygen.

For cutting steel, the metal to be cut is heated by a flame to ignition temperature and a jet of pure oxygen is directed on to the heated surface. The metal burns in the oxygen and the oxide debris is blown out of the cut; the nozzle is moved steadily in the required direction of cut, Fig. 10.2. When cutting metals which have refractory oxides, such as stainless steels or aluminium, an abrasive powder is fed to the work zone to cut away the oxide; modern practice favours plasma or laser cutting of such materials.

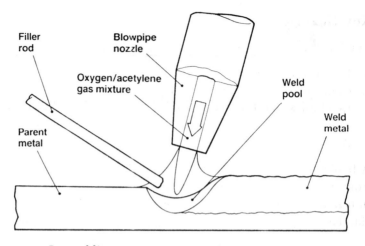

10.1   Gas welding.

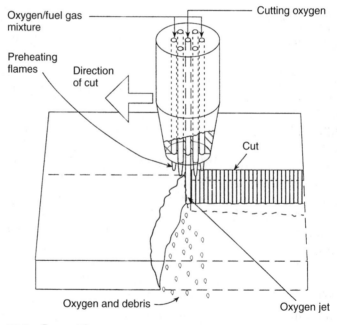

10.2   Gas cutting.

Key Hazards

Key hazards that are encountered are as follows:

- Explosion and fire
- Compressed gases (unexpected release of pressure)
- Eye injuries and burns
- Fume
- Accumulations of gas (e.g. by leakage, or unburnt gases).

A booklet giving guidance for gas welding and cutting is available from the HSE[123] in the UK. The American Welding Society publication Z49.1[124] gives advice for American users. The original Federal standards are given in the list of references at the end of the book.[125-127]

Explosion and Fire

The key elements in the prevention of explosion or fire are the correct storage, transport and use of the compressed gases used in welding.[128] In addition, using the correct equipment and procedures when using the naked flames will reduce the risk of accidents in the workshop. Serious or even fatal accidents have occurred owing to the catastrophic failure of equipment that was rated below its service pressure, as a result of high pressure oxygen being released onto organic material, and catastrophic failure of acetylene cylinders in fire.

While the precautions below will help to reduce the risk of an explosive or flammable mixture being released into the workplace, there is always the risk that a small leak may go unnoticed. For this reason, there should always be adequate ventilation during welding operations, the gases should be stored outside whenever possible and gas cylinders should not be taken into poorly ventilated areas or confined spaces. Welding on tanks, drums, etc, that have contained flammable material is discussed in Chapter 24.

Storage and transport of compressed gases

Chapter 4 described the properties, the general requirements for the storage of compressed gases at premises and the regulators and gauges that should be fitted. As little as 2.5% acetylene in air can burn, and air/acetylene mixtures in the range 2 to 82% acetylene are explosive. Mixtures of propane and air in any concentration between 2 and 9% are liable to explode. Special care is needed as propane is heavier than air and can collect in sumps, tanks, etc. If it enters

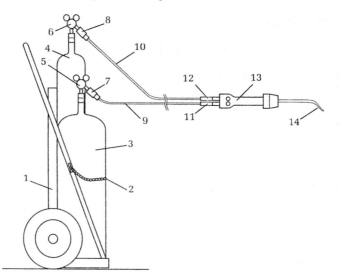

10.3 Mobile welding set: 1-Wheeled trolley; 2-securing chain; 3-acetylene cylinder; 4-oxygen cylinder; 5-acetylene regulator; 6-oxygen regulator; 7-acetylene flashback arrestor; 8-oxygen flashback arrestor; 9-acetylene hose; 10-oxygen hose; 11-acetylene non-return valve; 12-oxygen non-return valve; 13-torch body; 14-nozzle (interchangeable).

a drainage system, it can follow the fall of the pipes or gullies and travel distances of the order of kilometres until it reaches a source of ignition. If this hazard is present it is preferable to use acetylene, which is lighter than air and will rise away from the work area.

Welding and cutting sets

All gas equipment should be purchased from suppliers who are prepared to undertake that it has been manufactured to comply with recognised standards or other appropriate specifications.

To take advantage of the independence of gas processes from electricity supplies, particularly when applied to repair, maintenance and installation work on site, the necessary equipment is usually assembled on a wheeled trolley, Fig. 10.3. Mechanised gas cutting equipment, although using the same basic components, is normally permanently installed. In either case, the gas cylinders should be sufficiently far away from the flame so that sparks do not reach them. In addition, they should be sited so that valves can be reached quickly in the event of an emergency.

Table 10.1. Hose colour code

Gas	Hose colour (UK)	Typical hose colour (USA)
Acetylene	Red	Red
Hydrogen	Red	Red
Propane and LPG	Orange	Orange
Oxygen	Blue	Green
Air or inert gas	Black	Black

Hoses and materials for pipework

The blowpipe should be connected to the regulator by means of the correct type of hose.[129–131] Hose is generally colour coded, as specified by the body having local jurisdiction. Table 10.1 gives the typical colour coding.

The hose should be securely attached to the blowpipe, the regulator and any connection by means of suitable, non-reusable ('single shot') hose-clips.[132–134] The use of pieces of pipe with twisted wire, or the worm-driven clips, must be avoided, since they can fail without warning. New hose is generally supplied complete with the correct types of fitting. When it is necessary to couple together two lengths of hose, the special couplers available should be used. The hose should be inspected frequently for cuts, burns, abraded areas or cracks and, if these are found, the defective hose should be renewed.

Hose should not be longer than is needed and should be protected from heat, mechanical damage, sparks, slag, oil and grease. Hose may be taped together to aid good housekeeping, but not more than one third of its length should be obscured by tape. Fire in a coiled hose is difficult to extinguish, so hose should not be coiled during operation, especially around cylinders, regulators or the handle of the trolleys.

Hose made from thermoplastic material is not generally suitable for welding, because it cannot resist the hot particles. However, for miniature work, such as jewellery, it may prove impossible to find any reinforced rubber of the correct dimensions. Fortunately, the conditions for such work are not so severe.

Where lengths of hose are made up with end fittings supplied separately, particular care must be taken to ensure that right-hand thread fittings, with plain hexagon nuts, are always used for the oxygen hose, and left-hand threads, with nicks on the corners of the hexagons, for fuel gas. Fittings to be used for gas supplies should be purchased from a reputable equipment supplier. No high pressure fitting exposed to cylinder pressure should be made except by those with the necessary expertise and facilities for pressure testing.

If fittings have to be made up for the acetylene supply it is most important to ensure that copper, copper-rich or silver-rich alloys are not used. Copper or silver in contact with acetylene are liable to form dangerously explosive acetylides. Only metal containing less than 70% copper should be used. If fittings are silver-soldered, no more than 0.3 mm width of filler metal should be exposed at the joint and the filler should not exceed 21% copper and 43% silver.[56] Light metals, such as aluminium are not suited to oxygen service. The recommended materials include copper and its alloys and ferrous alloys.

Unions, nuts and connectors should be inspected. Faulty seats are liable to lead to leaks and should be discarded. Valves and fittings for all purposes should be kept scrupulously clean, and care should be taken to make certain that no grit or other foreign matter is allowed to remain on them. Attention to this small point will save much trouble arising from leaks and will prevent the build up of any dangerous concentrations of gas. In addition, for oxygen service there must be no trace of oil, grease or other organic material – if lubricants are necessary, they must be of a type approved for high pressure oxygen service. Care should be taken to avoid leaving any swarf when assembling the system. The assembled equipment should be checked for leaks with a 0.5% detergent solution.

Should a valve, regulator, or any other piece of equipment become frozen, it should be thawed out by means of hot (but not boiling) water; no other method should be employed to thaw equipment.

Blowpipes or torches

These are designed to burn the fuel gas with air or oxygen. They are supplied with gas inlet connections, a handle and separate control valves for each gas. Internally there is a mixing chamber for the gases and the flame emerges from the nozzle. In a torch designed for gas cutting, there is an additional channel for the cutting oxygen.

Blowpipes commonly used with LPG in plumbing and roofing are 'air-aspirated', that is, they draw air from the atmosphere to mix with the fuel just upstream of the nozzle. Blowpipes may be of the mixer type, requiring equal gas supply pressures, or injector, with oxygen supplied at higher pressures than acetylene, see Figs. 10.4 and 10.5.[135]

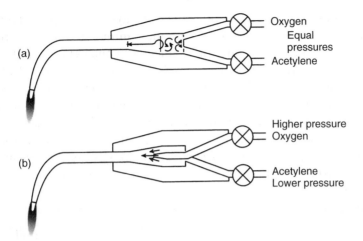

10.4   Blowpipe principles (a) mixer, (b) injector.

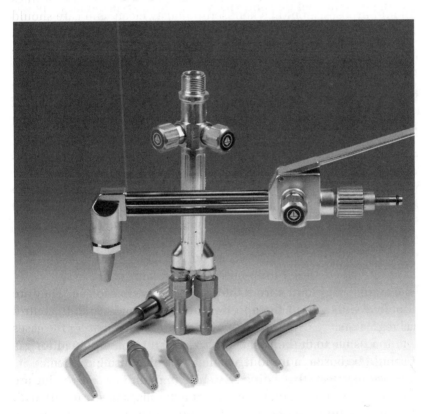

10.5   Torch body, with welding and cutting attachments (courtesy Messer-Griesheim).

The user should check that the blowpipe to be used is suitable for the gases to be used and suited to the temperature, pressure and gas flow.

Nozzle tips should only be cleaned with the reamer supplied for the job; a drill will enlarge the hole and adversely affect the flame. Blowpipes should be regularly inspected to make sure they are undamaged, in good working order, and are not leaking, especially at the valves; a flame here could set the welder's clothing alight. If repair is needed, or in any event at regular intervals, they should be dismantled and cleaned, preferably by the makers or a specialist.[136]

Non-return valves

It is strongly recommended that non-return valves are fitted to both the gas lines at the blowpipe inlet to reduce the risk of oxygen flowing into the fuel line and vice versa. Older style devices, known as 'hose check valves' or 'hose protectors' rely on a floating plate to stop the flow. These are ineffective at low pressures and should be discarded. Non-return valves complying with BS EN 730 or equivalent should be used.[137] Non-return valves are not needed with air-aspirated blowpipes.

Flame or flashback arrestors

Should a backfire be propagated back into a hose from the blowpipe it is possible for it to continue back towards the regulator. If it reaches the acetylene cylinder it may internally fire it and eventually it could explode. A flame arrestor contains an element that rapidly quenches a flame. These devices also frequently incorporate a pressure or temperature actuated cut-off valve in order to cut off the gas supply. It is strongly recommended that a flame arrestor with a cut-off valve is fitted to the pressure regulator outlet of all acetylene cylinders, or outlet from the acetylene distribution system. It is also good policy to fit them to the outlets of oxygen and other fuel gas regulators.[137–139]

It is possible to fit the flashback arrestor at the blowpipe inlet, but it should be borne in mind that this leaves the hose unprotected in the case of a leak that is inadvertently ignited. For long hoses (in excess of 3 m), it is recommended that a flashback arrestor with cut-off valve is fitted at the regulator outlet and a simple flashback arrestor is fitted at the blowpipe. Flashback arrestors are not necessary for air-aspirated blow pipes.

Working practices

To prevent fire in the workplace, welding should be carried out in a safe location. If this is not possible, then all combustible material within about 10 m should be removed, or covered with suitable guards or blankets (see Fig. 3.1 for a sample material). Attention should be paid to any area where sparks, etc, can escape through cracks into other areas. Ventilation should be good, to disperse any unburnt gases. A fire watch should be kept throughout the work, and for at least an hour afterwards. A suitable fire extinguisher(s) should be kept near at hand.

Unexpected Release of Pressure

If the regulator fails, the gas is released suddenly, but failures of regulators that have been purchased to the correct standard and selected correctly for the service are rare. However, there are two common sources of accidents in handling gas cylinders. A typical oxygen cylinder is filled to a pressure of 240 bar, and weighs 60 kg or more. The regulator has a relatively narrow stem where it joins the cylinder. If a gas cylinder is knocked over with the main valve open, there is a real risk of the regulator shearing off. This will cause the cylinder to become rocket-propelled as the gas escapes. There is potential for very serious, or fatal injury. Hence, it is imperative that cylinders are kept tethered, either in a trolley, or to an immovable object all the while they are in use. When they are to be transported, it is essential that the valve is closed and it is preferable that the regulator is removed altogether.

The importance of careful handling of compressed gas cylinders cannot be too strongly emphasised and it is most dangerous to use them as supports for the work or to allow them to remain near furnaces or other sources of heat, since they may overheat and vent or explode.

Eye Injuries and Burns

Precautions must be taken to prevent burns to the eyes and exposed parts of the body. These may occur as the result of spattering of incandescent metal particles and from flying slag particles. The intense radiation from the flame and incandescent metal in the weld pool can cause considerable discomfort to the operator and others

10.6   Eye protection for gas welding (photograph courtesy of Racal Safety).

Table 10.2. Filter scale numbers to be used for gas welding (UK)[98]

Welding	$q \leq 70$	$70 < q \leq 200$	$200 < q \leq 800$	$q > 800$
Welding of heavy metals (e.g. steels)	4	5	6	7
Welding with emittive fluxes (notably with light alloys)	4a	5a	6a	7a
Cutting	$900 \leq q \leq 2000$	$2000 < q \leq 4000$	$4000 < q \leq 8000$	
Oxygen cutting	5	6	7	

q = Flow rate of acetylene in litres per hour ($l\,hr^{-1}$).
a = Scale numbers for welding with emittive fluxes.

in the vicinity of the operation. Eye protection should be worn, with a filter to reduce the harmful radiation; it is recommended that the eye protection has side shields, see Fig. 10.6.

Table 10.2 shows the recommended filters for eye protection according to the UK Standard.[98] According to the conditions, the next greater or smaller number may be used.

Table 10.3 shows the recommended filters for eye protection according to the US standard.[101] It is important to choose a filter that absorbs yellow.

Table 10.3. Filter scale numbers to be used for gas
welding and cutting (USA)[101]

	Plate thickness		
Gas welding			
Light	<1/8″	<3.2 mm	4 or 5
Medium	1/8 to 1.2″	3.2 to 12.7 mm	5 or 6
Heavy	>$\frac{1}{2}$″	>12.7 mm	6 to 8
Oxygen cutting			
Light	<1″	<25 mm	3 or 4
Medium	1–6″	25–150 mm	4 or 5
Heavy	>6″	>150 mm	5 or 6

A good general standard of illumination at the work is essential
in welding shops and booths. The operator's body and clothing must
be adequately protected from sparks, flying particles of incandescent
metal or slag. This will probably consist of a boiler suit, gloves and
a cap. No oily or greasy clothing of any kind should be worn. Clothing should be made from cotton or wool. Some fabrics are available
that have been treated to render them flame retardant.

Articles which have been welded will be very hot on completion.
It is recommended that these should always be clearly marked HOT
to warn other employees who may have to handle them. The
marking should be removed when the article is cool enough to be
handled without injury if the most effective protection is to be
achieved. In practice, a reasonable rule would seem to be that every-
thing on a welding bench should be treated as hot, and that articles
which are not in areas protected by ropes or barriers should be individually marked. Small piece parts may be placed in a marked container. If it is thought that an article may be hot it may be approached
cautiously with the back of the hand, which is sensitive to radiation
from a very hot object; if done carefully, it should be possible to tell
whether the item is hot without getting burnt.

Fume Risks

Some indication of the fume concentrations is given in Table 5.4.
The quantity of fume from the metal and filler are relatively modest
for gas welding, but larger for gas cutting and gouging. Nevertheless,
if welding or cutting toxic materials, precautions must be taken.

Good ventilation must always be provided for gas welding. The
heat produced by prolonged contact of the acetylene flame with a

large mass of metal leads to the formation of oxides of nitrogen. In confined spaces dangerous concentrations can build up. Fatal accidents have occurred owing to the inhalation of excessive amounts of oxides of nitrogen in preheating. Unfortunately, the person affected is unaware that an overdose is being received. Good ventilation in these cases is essential.

The fumes given off when welding and cutting parts which have been galvanised, lead-coated, or otherwise treated, may be injurious to the operator, and special precautions must be taken. Local exhaust ventilation should be used. If this cannot guarantee the safety of the operator then a respirator may be required. The powder cutting process used to cut stainless steel and non-ferrous metals also requires special precautions. For mild steel, in a normal workshop environment good general ventilation will normally be sufficient for gas welding.

Accumulations of Gas

Normal air contains only some 21% oxygen; the remainder, mainly nitrogen, takes no part in most combustion reactions and so slows down the burning by simple dilution. If the oxygen content is increased, burning intensity and speed is increased, normally non-flammable materials may burn and oil or grease may catch fire spontaneously. Oxygen may be released into the air by leaks in equipment, by supplies being left on or by excessive purging. In the normal operation of the flame cutting process about 30% of the oxygen supplied is released unconsumed to the atmosphere. Gas cutting should never be undertaken in a confined space without proper ventilation arrangements.

Note that, although fuel gases are treated with odorising agents, oxygen is odourless, and workers may not notice dangerous concentrations. It is very dangerous to search for gas leaks with a naked flame; only a weak (0.5%) solution of detergent in water should be used for this purpose. It is best to avoid the use of soap solution as this may react with oxygen when it dries out.

There have been a number of accidents caused by unburnt gas passing through the gaps in the preparation during preheating with oxypropane torches. It appears that this is a problem that arises if the gases can accumulate in the space behind the plates and is then subsequently ignited by the flame. Running the torch with a restricted oxygen supply makes the problem worse. An explosive mixture can be generated in only a few seconds.

Leaking hoses or equipment are always dangerous. For this reason, leakage checks should be done. It is preferable to leave the gas cylinders outside, if welding is to be done in a relatively enclosed area. Whenever the apparatus is left, turn off the gases at the cylinder.

Working Procedures

In common with the other welding and cutting processes, gas welding and cutting is quite safe if elementary precautions are taken.[123,139,140] All equipment should be operated in accordance with the manufacturers' recommendations. Elimination of danger from welding and cutting is more often than not a matter of the application of sensible precautions; carelessness can so easily lead to personal injury or damage to property. The equipment is often used by, for example, maintenance workers who have not received specific training; in such circumstances rigorous supervision and control of portable equipment is essential.

Workers should be trained to use the equipment correctly. This training should include information on how to select the correct equipment and how to check that it is working correctly. They should know how to assemble it correctly and check that it is free from leaks. They should be taught the correct method of lighting and using the flame and the correct method of shutting down. They should be given training in how to deal with the common emergencies.

Supply hoses should be arranged so that they are not likely to be tripped over, cut or otherwise damaged by moving objects, as a sudden jerk or pull on the hose is liable to pull the blowpipe out of the operator's hands, cause a gas cylinder to fall over, or a hose connection to fail.

It is important to purge the gas lines, one by one, before using the equipment, to avoid the formation of explosive mixtures in them. Before lighting the blowpipe, fuel gas and oxygen must be allowed to flow for a few seconds (or more for long lengths of hose) separately through the systems to the blowpipe tip, ensuring that each gas line (regulator, hose, etc) contains only its own gas, and not a mixture, regardless of the previous history of the equipment.

Pre-use Equipment and Area Checks

First check that all the equipment is of the correct type, rated to the correct pressures and of suitable materials. In particular check that the regulators are correct for the gas and the pressure, and that they are not damaged. Check the condition of the threads and the sealing surfaces, check for oil or grease contamination. Check that the flash-back arrestors are correct for the gas and the pressure, and that their threads are in good condition. Check hose and hose assemblies for damage. Check all connections for leakage at the working pressure using a detergent solution.

Ensure that the gas cylinders are placed so that they are not going to be showered with sparks or spatter, or where they may become part of an electric circuit. Ensure that the hoses are placed so that they will not suffer damage. Examine the blowpipe nozzle and inlet seatings for damage. Leak test all joints at the working pressure. Ensure that the work area is free from combustible materials. Sparks can ignite materials 10 m away, or more.

The standard procedure for lighting-up is as follows:

1 Carry out the pre-use equipment checks.
2 Ensure that all valves are closed and the regulator pressure adjusters unscrewed.
3 Open the oxygen cylinder valve; limit opening to half a turn, unless the cylinder supplier advises otherwise for a 'soft seat' valve (typically two turns).
4 Screw in the oxygen regulator adjuster to approximately the correct outlet pressure.
5 Open the oxygen blowpipe valve half a turn and allow the gas to purge the hose. Set the pressure finally, then close the valve at the blowpipe.
6 Repeat steps 3 to 5 for the acetylene supply and light the fuel gas immediately, preferably with a spark lighter (not matches, cigarette lighters or welding arcs).
7 Open the oxygen valve and adjust it and the fuel gas to obtain the type of flame needed for the job in hand.

When the welding run is ended:

8 Turn off the torch acetylene valve, then the oxygen. The torch can be relit from step 6.

When the equipment is to be left unattended, after step 8:

9 Close the cylinder valves.
10 Open the acetylene torch valve to release gas in the regulator, then close it.
11 Repeat step 11 for the oxygen.
12 Unscrew the regulator adjusters.

Contingencies

Backfire

Backfire is when a flame burns back into the blowpipe, perhaps accompanied by a bang. This is caused by the blowpipe being too close to the workpiece or by a blockage in the nozzle. The flame may go out or may re-ignite. There is insufficient pressure for the nozzle used or the nozzle is overheated. Shut off the valves at the blowpipe, oxygen first, then fuel. Shut the valves at the cylinders. Cool the blowpipe, using water if necessary. Check for damage especially to the nozzle.

If any heating of the hoses is apparent, or in the event of any other problem such as leaking gas catching fire, turn off the cylinder valves, acetylene first; this is helped by the limited valve opening in steps 3 and 6 above. If any resulting fire cannot be put out straight away – evacuate. If the acetylene cylinder starts to get hot or to vibrate, evacuate and call emergency services.

Flashback

This can occur if there is a flammable mixture in the hoses when the torch is lit. If not stopped, it can go through the regulator into the cylinder and can trigger the decomposition of the acetylene. Flashback may be caused by reverse flow of oxygen into fuel or vice versa producing an explosive mixture in the hose, or if the lines have not been purged. It can be serious because it can cause the cylinder to explode. Use the correct lighting up procedure, purge the lines. To protect the system ensure non-return valves are fitted and that there are flashback arrestors on both gas lines (at both ends if they are long). Ensure the gas pressures are correct and maintain equipment in good condition.

Flashback may also result from dipping the nozzle tip into the molten pool, mud or paint, or from any other stoppage at the nozzle; the obstruction so formed causes the oxygen to flow back into the acetylene pipe and communicate ignition back towards the generator or cylinder. Any particles of slag or metal that become attached to the tip should be removed and if the blowpipe tip becomes hot when working in a confined space or close to a large mass of hot metal it should be cooled frequently by immersion in a bucket of water after extinguishing the flame.

Flashback arrestors may be damaged when they are exposed to flashback. Consult the supplier and replace if necessary.

Overheating cylinders/fire

Should an acetylene cylinder become heated accidentally, or become hot as a result of excessive or severe backfire from the use of faulty equipment, the gas manufacturers recommend that it be dealt with promptly as follows:

> 'Shut valve, detach regulator, remove cylinder outdoors at once, spray with water to cool, keep cool with water. Leave outdoors. Advise suppliers immediately, quoting cylinder number where known'.

If fire should break out, the first actions should be to raise the alarm, in order to evacuate personnel from the area, and to call the emergency services. Subsequently, decisions will need to be made about whether to remove the cylinders and how to manage them. A contingency plan should be drawn up and personnel trained to take

Table 10.4. Maintenance

Item	Annual inspection	Replacement intervals
Regulators	Functional tests to ensure internal components are operating correctly	5 years, or at the supplier's recommendation
Flashback arrestors	Reverse the flow to check the operation of the internal components. Check the flow in the correct direction with the cut-off valve tripped (if pressure-sensitive type)	5 years, or at the supplier's recommendation
Hoses and non-return valves	Reverse hose to ensure the correct operation of the valve. Bend the hose to a tight radius to see whether the reinforcement is visible	Determined by the operating conditions
Blowpipes	Test the valves for their function. Blank the exits and test for leaks	Determined by local conditions

appropriate action. There are documents giving further advice on the handling of acetylene cylinders in a fire.[128,141]

Maintenance

Annual maintenance of gas welding equipment should be carried out by a person who has sufficient practical experience of the equipment and a theoretical knowledge of the functioning of the equipment.[128] They should know the properties of the gases used and the potential problems that may occur. Some suggestions are made in Table 10.4.

11

Arc Welding and Cutting

The Electric Arc

In arc welding the heat source is an electric arc, which is formed either between a non-consumable electrode or a consumable electrode and the workpiece. An alternating current (AC) or direct current (DC) power supply is connected to an electrode and to the workpiece; an arc is struck between electrode and work, melting the work to make the joint. A consumable electrode, if used, will also melt and add filler metal to the weld pool. If a tungsten electrode is used, its melting point is so high (about 3200° C) that it does not melt appreciably – a 'non-consumable electrode'. The joint may be formed by melting only the parent material – 'autogenous welding' – or from a 'filler rod' melted into the joint. There are standards relating to arc welding and advice booklets.[124,142]

Consumable Electrode Processes

Manual metal arc welding

Manual metal arc (MMA) (UK) welding processes are consumable electrode processes that include the following close variants and alternative names: stick welding, electric arc welding, shielded metal arc welding (SMAW) (USA), touch welding and gravity welding (with simple mechanisation).

In these processes, the electrode is usually of similar composition to the work and heat from the arc melts the end of the electrode as well as the work. Metal is transferred across the arc from the electrode to the work to form part of the weld. The electrode is advanced to maintain a steady arc length. The various consumable electrode processes are further distinguished by the means adopted to shield

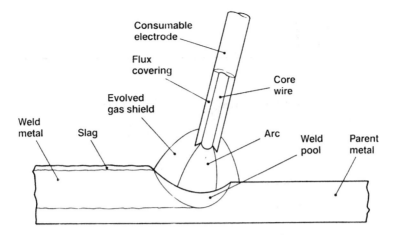

11.1   MMA welding.

the weld region from reaction with the atmosphere. In MMA welding, this is gas which has been released from the flux coating of the electrode.

The electrode consists of a core wire covered with flux which is made from powdered minerals and sometimes metal powders to alter the composition of the weld metal. The flux reacts during welding to form a shielding gas to protect the arc zone and the molten weld pool, and to form a slag to protect the cooling weld metal. The slag also takes part in metallurgical reactions with the molten weld metal. Current is fed into the electrode at the far end, usually via a hand-held electrode holder, see Fig. 11.1.

Gas-shielded welding processes

Gas-shielded welding processes include metal inert gas (MIG) welding, metal active gas (MAG) (UK), CO_2 welding, gas-shielded metal arc welding, semi-automatic welding and gas metal arc welding (GMAW) (USA).

The arc and the weld zone are shielded by gas supplied from a cylinder; the gas may be either carbon dioxide, argon, helium or a mixture of these, with or without small additions of oxygen. The solid wire electrode, supplied on a reel, is fed in by a motor to maintain a constant arc length. Note that many steel wires are supplied lightly copper plated to help prevent rusting, see Fig. 11.2.

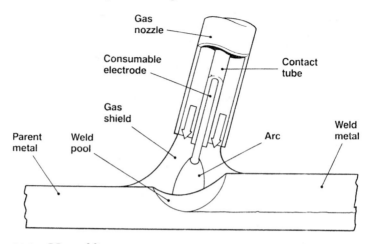

11.2   CO$_2$ welding.

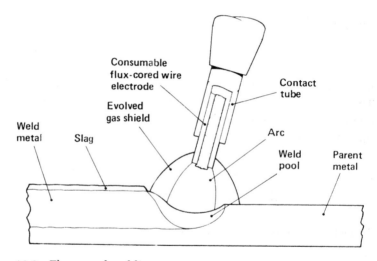

11.3   Flux-cored welding.

Flux-cored welding

Flux-cored welding is also known as inner shield welding, self-shielded welding and flux-cored arc welding (FCAW).

This is a variant of the above gas shielded processes, in which the electrode wire is hollow. The cavity is packed with a flux core, which generates gas to provide shielding, either on its own, or with further gas from a nozzle as in the previous process, see Fig. 11.3.

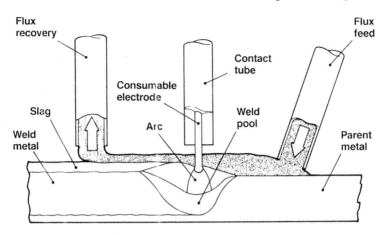

11.4   Submerged-arc welding.

Submerged-arc welding (SAW)

The arc is submerged beneath a covering of granulated flux, which protects the arc zone and the weld from atmospheric attack, and may take part in metallurgical reactions with the molten weld pool. The electrode is a bare wire which is automatically fed to maintain a constant arc length, see Fig. 11.4. The wire is commonly copper plated.

Non-consumable Electrode Processes

Tungsten inert gas welding

Tungsten inert gas (TIG) welding (UK) uses a non-consumable electrode and has variants and alternative names that include: gas-shielded tungsten-arc welding, gas tungsten arc welding, GTAW (USA), argon arc welding, heliarc and heliweld.

A tungsten electrode is used and the electrode, filler and weld metal are protected from the atmosphere by a shield of inert gas, usually argon or helium. If required, extra metal to form the weld may be added in the form of filler wire. The arc may be started and maintained by superimposing a high voltage on to the main welding supply at a high frequency or as a train of pulses, see Fig. 11.5.

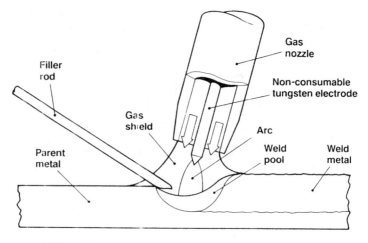

11.5   TIG welding.

Cutting and Gouging Processes

Oxygen arc cutting; oxyarc cutting

An arc is struck between the workpiece to be cut and an electrode covered in flux, similar to an MMA electrode, but with a tubular core; oxygen is injected through the core. When the work heats up to a sufficient temperature, it starts to burn in the oxygen and the electrode is moved in the required direction of cut, see Fig. 11.6.

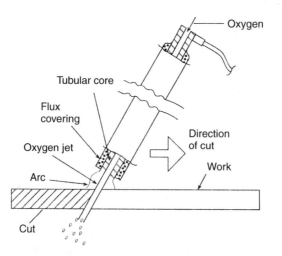

11.6   Oxygen arc cutting.

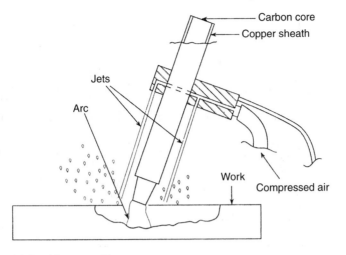

11.7   Air arc cutting.

Air arc cutting; arcair cutting; arcair gouging; air carbon arc cutting and gouging

The electrode is mainly carbon, with a thin copper coating to reduce electrical resistance. When the arc has melted the work, compressed air jets are turned on to blow away the molten metal and the electrode is manipulated to cut or gouge as required, see Fig. 11.7.

Other Arc Processes

Stud welding; arc stud welding

To weld a stud to a flat surface:

1   The stud is placed in contact with the workpiece.
2   Current is initiated and the stud withdrawn to start an arc.
3   The arc extends to melt the end of the stud and forms a weld pool.
4   Current is switched off and the stud forced into the weld pool to complete the joint.

For steels, the ceramic ferrule provides enough shielding for the few seconds arcing time; for aluminium, inert gas shielding is required, see Fig. 11.8.

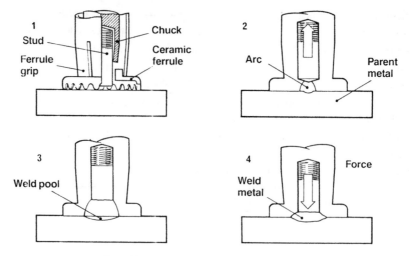

11.8   Stud welding.

Key Hazards

Key hazards are:

— radiation (non-ionising)
— fume
— electricity
— fire
— noise
— spatter
— eye injuries caused by deslagging
— magnetic fields
— ionising radiation (from thoriated tungsten electrodes).

The relative risks from these vary between the processes. For instance, submerged arc welding produces a minimal risk of exposure to arc radiation or fume.

Radiation

Risk to the eyes

If the eyes are exposed to the light of the arc, even for quite a short period, arc eye may develop; this can be avoided by using a head shield fitted with a suitable filter and by avoiding stray flashes from other welding arcs. Filters may be either fixed transmittance or switchable.[98,101,102] Switchable filters react to the light of the arc and automatically darken to the chosen shade. This can be of advantage, since the welder has a clear view of the work before the arc is struck and does not have to pause to lower the helmet.

There is an 'urban myth' regarding the wearing of contact lenses and a risk that the lens may get welded to the eye. This story is unfounded. Welders and others in the welding shop may wear contact lenses if they wish. If, however, some grit or a chemical gets in the eye, the contact lenses should be removed immediately and first aid sought. Of the processes discussed in this chapter, sub-merged arc welding has little or no risk of arc eye, since the arc is rarely seen.

The filters suggested for arc processes are in the tables below. Table 11.1 summarises the British Standard and Table 11.2 the USA Standard.

Adequate screening to protect workers in the vicinity is essential. Where the work is carried out at fixed benches or in welding shops, permanent screens should be erected.

Welders should use screens or curtains to protect others in the vicinity from the arc. These are conveniently lightweight screens mounted in frames, or curtains hung from frames. There should be space at the bottom to allow ventilation to be adequate through-out the workshop. Screens are available to a British Standard. Material used for welding curtains should be fire resistant, see Fig. 11.9.[96,97]

Transparent tinted or translucent material is now available which affords some view of the work from the outside, helping to avoid accidents arising from lack of visibility, and offering reasonable resistance to heat and fire. In addition, the tinted sheet is claimed to meet similar standards to normal filters, although a lighter shade will suffice than that needed for close observation of the weld pool. In the case of untinted translucent curtains, that is those with a 'frosted finish', scattering of the light alone over-comes the hazard. The rippled surface adopted to cause scattering of trans-mitted light also prevents mirror-like reflections, which can distract the welder if tinted curtains with a shiny surface are used.

If any person is exposed to a flash, they should leave the area. If they experience the effects of arc eye (a feeling of grit in the eyes, and pain) they should have their eyes checked by a physician to rule

Table 11.1. Filters suggested for arc welding and gouging (UK)[98]

Usage	Current range (A)	Filter
MMA	<40	9
	40–80	10
	80–175	11
	175–300	12
	300–500	13
	>500	14
MIG Heavy metals	<100	10
	100–175	11
	175–300	12
	300–500	13
	>500	14
MIG Light alloys	<100	10
	100–175	11
	175–250	12
	250–350	13
	350–500	14
	>500	15
TIG	<20	9
	20–40	10
	40–100	11
	100–175	12
	175–250	13
	>250	14
MAG	<80	10
	80–125	11
	125–175	12
	175–300	13
	300–450	14
	>450	16
Arc-air gouging	<175	10
	175–225	11
	225–275	12
	275–350	13
	350–450	14
	>450	15

out the presence of any foreign body. The arc eye condition will recover spontaneously.

Painting arc welding booths

The use of black paints for the inside of welding booths has become a common practice, but a lighter shade is preferable, because it promotes better all round illumination. Paints using titanium dioxide

Table 11.2. Filters for use with arc welding processes (USA)[100,101]

Operation	Electrode size	Current	Minimum shade	Suggested shade
SMAW	<3/32", 2.5 mm	<60	7	–
	3/32"–5/32", 2.5–4 mm	60–160	8	10
	5/32"–7/32", 4–6.4 mm	160–250	10	12
	>8/32", >6.4 mm	250–550	11	14
GMAW		<60	7	–
FCAW		60–160	10	11
		160–250	10	12
		250–500	10	14
GTAW		<50	8	10
		50–150	8	12
		150–500	10	14
Arc carbon	(light)	<500	10	12
	(heavy)	500–1000	11	14

or zinc oxide in the pigment do not reflect ultraviolet light and are generally suitable.[93]

Viewing distances

There are no definitive guides to the distance from which it is safe to view arcs without protection. However, some information is given in the literature[143] and an extract is reproduced as Table 11.3. Some experienced welders have commented that they have experienced arc eye after lesser exposures than this; possible sources of this discrepancy include:

– A synergic effect between ultraviolet (UV) exposure and welding fume, producing an effect in combination which is not apparent at the same level of either separately
– Failure of memory – the incidents recalled by the welders were all several years old
– Inappropriate exposure standard
– Experimental inaccuracies – a study of the full reports makes this seem the least likely.

Therefore these figures should be taken only as a guide and it may be prudent to increase the distances. Best practice is always to shield the arc. When light radiates from a source it obeys an inverse square law, so that doubling the distances will reduce exposure by a factor of four.

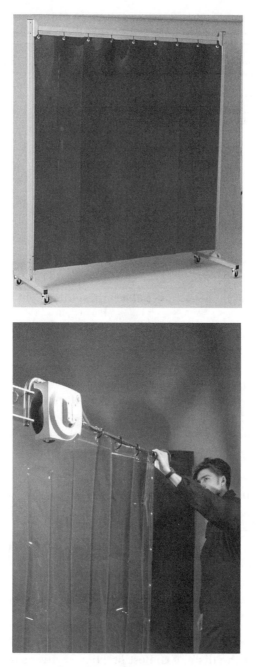

11.9   Adjustable welding screens (photographs courtesy of Tusker, Safety First Manufacturing Co and Nederman).

Table 11.3. Distance from arc at which ultraviolet radiation is reduced to the USA daily threshold limit value for various exposure times, from data by Lyon *et al.*[143]

Process	Parent metal	Shielding gas	Current (A)	Distance (m) for 1 min	10 min	8 hr
MMA	Mild steel	–	100–200	3	10	70
MIG/GMAW	Mild steel	CO_2	90	0.9	3	20
			200	2.1	7	50
			350	4	13	90
Flux cored wire	Mild steel	CO_2	175	1.2	3.5	24
			350	2.2	7	50
	Mild steel	95% Ar, 5% O_2	150	3	9	65
			350	6.5	20	140
	Aluminium	Ar	150	3	10	70
			300	5	17	110
	Aluminium	He	150	1.3	5	35
			300	3	10	70
TIG	Mild steel	Ar	50	0.3	1	7
			150	0.9	3	20
			300	1.6	5	40
	Mild steel	He	250	3	10	70
	Aluminium	Ar	50 AC	0.3	1	7
			150 AC	0.8	2.7	18
			250 AC	1.3	4	80
	Aluminium	He	150 AC	0.7	3	20
Plasma arc welding	Mild steel	Ar	200–260	1.6	5	33
		85% Ar, 15% H_2	100–275	1.7	5.5	40
		He	100	2.9	9	65
Plasma arc cutting (dry)	Mild steel	65% Ar, 35% H_2	400	1.3	4	30
			1000	2.5	8	55
Plasma arc cutting with water injection	Mild steel	Nitrogen	300	3.2	11	75
			750	1.8	5.5	40

Effect of arc radiation on the skin

Arc radiation can cause severe burns to any part of the skin that is exposed. Thus the welder must be protected by clothing that is opaque to UV light. This will usually consist of a welding helmet or visor, cap, boiler suit, gloves and shoes. The materials must not readily melt or catch fire.[94]

In both gas-shielded metal arc and TIG welding processes, more UV light is emitted than in manual metal arc welding at the same current, so welders should take extra care that their skin is fully protected, for example at the back of the neck.

Welding booths and portable screens should be checked regularly to ensure that there is no damage which might result in the arc affecting persons working nearby.

Table 11.4. Relative fume levels in arc welding

Process	Material to be welded	Probable level of fume in relation to the exposure limits	Constituents of special concern
MMA	Mild and low alloy steels	Usually higher	Particulates from the base material, electrode and flux (possible fluorides)
	Stainless steel	Usually higher	As above, but the fume may also contain significant quantities of Cr(VI)
	Aluminium	Higher	Aluminium and aluminium oxide particulate. Could also contain significant quantities of ozone
	Other alloys	Higher	Consult the manufacturer's data sheet
MIG/MAG	Mild and low alloy steels	Higher	Particulates from the base material and electrode
	Stainless steels	Higher	As above, but also ozone
	Aluminium	Higher	Aluminium, aluminium oxide and ozone
	Other alloys	Higher	Consult the manufacturer's data sheet
FCAW	Mild and low alloys steel	Higher	Particulates from the base material and flux
	Stainless steels	Higher	Cr(VI) is likely to be present
TIG	Mild and low alloy steels	Lower	Shielding gas
	Stainless steels	Gases may be higher	Shielding gas and ozone
Arc cutting and gouging	Any	Very high levels	When cutting or gouging alloys containing chromium it is probably present as Cr(VI)

Fume

Fume generation and the general principles for its control were discussed in Chapters 5 and 6. Table 11.4 expands on the information that was contained in Table 5.4.

Employers must assess the possible effects of the fume that might be produced in welding and take steps to control the production of fume and the exposure of their personnel below any exposure limits. These were given in Tables 5.1 and 5.2. Manufacturers of welding consumables produce data sheets,[76,42] which will often contain a typical fume analysis. The presence of airborne substances from other sources, like shielding gases, surface coatings, solvents, action of light on the atmosphere, must also be taken into account in the assessment. The number of welders working in the area, the volume of the area and its level of ventilation will all be important factors. It is important to pass on the information about the hazards to the welders.

Manual metal arc, flux-cored arc, metal inert gas and metal active gas welding processes can all produce significant levels of fume from the heating of the parent plate and any filler or flux. Gouging produces very large quantities of fume. Special care must be taken in welding non-ferrous metals and certain alloy steels.

When welding mild steel, MMA welding will generally produce levels over the exposure limit of $5\,\mathrm{mg\,m^{-3}}$. Thus local exhaust ventilation will be required even in a well-ventilated workplace. The fume will contain both the constituents of the parent plate and the flux, which may contain fluorides. In the USA, when using consumables containing fluorides, the label shown in Fig. 11.10 is displayed.

FCAW generally produces more fume than manual metal arc welding. For MIG welding, dip transfer produces lower fume levels than MMA, pulsed transfer produces intermediate levels of fume and spray transfer generates higher fume levels. TIG welding will produce less fume, generally below the exposure limit for mild steel.

In addition to fumes evolved from electrodes and fluxes, coated or otherwise treated metals may give off toxic fumes. The UV light from the arc, particularly a gas shielded arc, may form ozone or phosgene from solvents and the heat may form nitrogen oxides from the air. Ozone levels are likely to be high in welding aluminium using MMA, MIG/MAG or TIG processes, and MIG/MAG and TIG welding of stainless steel. Local exhaust ventilation will be needed in addition to good general ventilation.

Shielding gases such as argon, helium and carbon dioxide are released into the atmosphere and may build up where ventilation is restricted, or even in open topped tanks as they are generally heavier than air. This may lead to a risk of asphyxiation. Carbon monoxide and carbon dioxide may form from the action of heat on the welding flux or slag.

Warning – contains fluorides.

Protect yourself and others. Read and understand this information.
Fumes and gases can be hazardous to your health.
Burns eyes and skin on contact. Can be fatal if swallowed.

- Before use, read and understand the manufacturer's instructions, the manufacturers' safety data sheet and your employer's safety practices.
- Keep your head out of the fume.
- Use enough ventilation.
- Avoid contact of flux with eyes and skin.
- Do not take internally.
- Keep children away when using.
- See ANSI Z49.1.

First Aid: if contact in the eyes, flush immediately with water for at least 15 min. If swallowed induce vomiting. Never give anything by mouth to an unconscious person. Call a physician.

Do not remove this information.

11.10   Warning label for brazing and welding fluxes containing fluorides.

It should not be overlooked that the simple matter of arranging the work in such a way that the welder does not lean over the fume greatly reduces the potential exposure. Further information regarding fume can be gained from AWS publications, HSE publications and the Welding Institute booklet.[77–79]

Electricity

Electricity can cause electric shock, burns, explosions and fires. In the welding environment, power supplies are frequently fed from mains, sometimes at high voltages (>400 V). The welding circuit itself will run at a low voltage and high current, but in order to strike the arc, the welding set will generally have a relatively high voltage open circuit (of order 100 V), which can cause fatal shock, particularly when welding in a conducting environment, such as damp conditions or inside metal vessels. The added precautions for environments with increased hazard are covered in Chapter 24.

Arc welding equipment

There are several generations of arc welding set in use. Very old welding equipment had a connection internally between the welding circuit and the case of the welding set. It is recommended that this type of equipment is scrapped.

Welding sets are also in use that have single layer insulation between the windings of the transformer. The cases of these welding sets must be earthed (grounded) to provide protection against the possibility that a fault will arise that will deliver electricity at the supply voltage to the welding circuit itself. In addition, the workpiece should be earthed by an earth lead independent of the return lead, see Fig. 11.11. This allows any fault in the insulation between the primary and secondary windings of the transformer to flow to earth and blow the fuse, protecting the welder. The earth connection must be capable of carrying the full secondary current without damage; cable similar to that of the welding circuit is suitable. It is important not to confuse the earth lead with the welding return lead – they are separate items. This is discussed in an HSE booklet.[144]

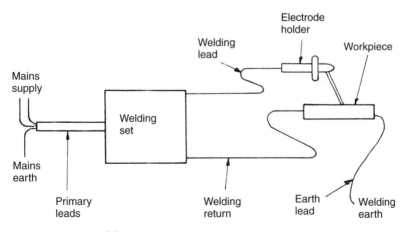

11.11   Welding circuit connections.

When purchasing arc welding equipment, ensure that it is suited to the purpose,[145–147] for instance if expecting to use the equipment in areas of higher hazard, ensure that the equipment has the correct level of ingress protection and the correct open circuit voltage (See Chapter 24).

Modern welding equipment has either double insulation in which there are at least two independent layers of insulation between the mains and any conductor that the user may come into contact with, or reinforced insulation. This type of equipment, when manufactured for the European market, is marked with the standard number EN 60 974-1. For this type of equipment, the use of a separate earth lead to the workpiece is not recommended. However, if an earth lead is required it should be connected as described above in this section. Because there are so many types of welding set in common use, it is important to check the rating plate of the equipment, which will give details of the standard to which it was made, the ingress protection (e.g. against water), whether it is double insulated, and whether it is suited to welding in locations of increased risk from electric shock.

It is important to guard against stray welding currents, because these can cause electrical hazards. Where the insulation of the return lead is faulty, these stray currents may be considerable and comparable to the welding current. Place the welding return lead as close to the point of welding as possible and preferably no further away than 3 m. Do not use metal rails, pipes or frames as part of the circuit, unless they are part of the workpiece.

Equipment should be maintained in accordance with the manufacturer's instructions, having taken into account the environment in which it has been used. Metal dusts can infiltrate into electrical equipment, which is likely to need cleaning at intervals.

Welding equipment is generally portable or transportable (i.e. on wheels), and thus should be part of the formal inspection and testing regime for such electrical equipment. It is recommended that it is inspected formally and the integrity of the earth and the insulation tested at least annually.[22]

Connection to the supply

The majority of arc welding equipment draws power from the mains electricity supply; engine-driven equipment, as used on site work, will be considered in the next section of this chapter.

With stationary transformers or motor generator sets it is recommended that a suitable switch fuse be mounted adjacent to the equipment so that it may be isolated from the supply main, if necessary. Portable sets with trailing cables should be provided with interlocked fuse-switch sockets and plugs at the supply end of the cable to give protection to the trailing cable as well as to the equipment. Should damage to the cable occur it can thus be isolated, as can the equipment, by means of the switch fuse. Use of the interlocked type of socket will ensure that the equipment is not plugged in or disconnected while on load. Where lengthy cables are in use, a switch or other means of cutting off the supply close to the operator may be required. If there is no switch on the set itself it may be desirable to fit one at that end of the cable. The switches should always be readily accessible, so that there is no delay in an emergency.

Engine-driven equipment

Engine-driven equipment should be sited so that no danger to health arises from the exhaust gases. Care should be taken to ensure that the plant is on a level platform and that the brakes have been applied or chocks fitted under the wheels to prevent movements from vibration. In very cold weather the unit should be positioned so that the prevailing wind is not blowing into the radiator. All mechanical guards should be in position and a satisfactory means should be provided to start the engine.

Engine maintenance should be carried out in strict accordance with the details given in the maker's handbook, the generator should be blown out from time to time, and all electrical connections checked and tightened. Particular care should be taken with regard to fuel leaks; all accumulations of fuel should be cleaned off and all connections made good. Care should be taken to avoid spilling the fuel when filling the tank.

Connecting the welding circuit

In service, the welding operator should check all external connections daily and should report any weaknesses, defects, etc, that are found. A periodic inspection should be carried out by a responsible person nominated for the task who should ensure that all connections are clean and tight, that they are correctly made, that the correct types and sizes of cable, earthing clamps, electrode holders, cable connectors, etc (Fig. 11.12) are being used, and who should particularly ensure that the earthing arrangements are satisfactory in all respects. Any connection which is hot to the touch after use should be dismantled, cleaned as necessary and re-assembled.

The current ratings of most arc welding equipment and accessories take account of the 'duty cycle', the percentage of time for which current is flowing, calculated on a total time of not more than 5 min. In much manual metal arc welding, the duty cycle is limited to some 35% by the need to set up parts for welding, to change electrodes, to chip slag, etc.

The size and construction of cables suitable for welding circuits is shown in BS 638 part 4.[148] Permissible currents for each type of cable at given duty cycles are tabulated. When applying these, note that it may be necessary to use larger cables than those indicated to reduce the voltage drop to an acceptable figure and that cables should be allowed free ventilation; for example, they should not be coiled up while in use.

Particularly with manual metal arc (MMA) or similar processes where the electrode holder is permanently live, it is necessary to consider where the circuit can be broken should a welder receive a shock and be unable to let go. Some patterns of connectors can be disconnected live in an emergency or it may be possible to switch off at the set itself or at the mains supply to it. The disconnection point should be readily accessible and reasonably close to the operator; 10 m is suggested as a reasonable distance but this will depend on local conditions.[149]

Accessories

Operators' hand shields and helmets, Fig. 11.12a and b, should be checked from time to time to ensure that there is no damage, such as burn holes, which would result in light being admitted or would provide a conducting path to the inside. A check should also be made that an unbroken filter of a suitable shade is fitted and that the cover sheet is present and reasonably free from spatter.[150]

Risk of electric shock

In welding, the voltage to earth may be high enough to cause fatal shock, if welders make contact with a live part with one part of

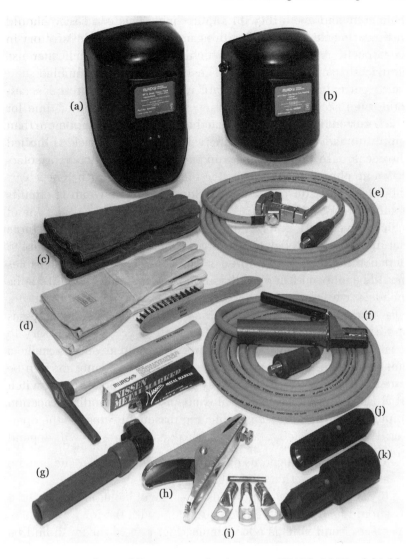

11.12 Arc welding accessories (courtesy ESAB). (a) Handshield,
(b) helmet, (c, d) protective gloves, (e) welding return cables
with screw clamp, (f) welding cable with electrode holder type
A, (g) electrode holder type B, (h) spring clamp for return or
earth connections to work, (i) connection lugs for crimping
to the cable, (j) cable socket and (k) cable plug (for joining
cables).

their body and an earth with another part. This can easily happen in damp conditions, if their gloves are faulty or if their head makes contact with an earthed object such as a steel girder. The hazards of electric shock are known to be less with a DC supply, but some welding sets have a substantial AC ripple, rendering them less safe than true DC.

Where welding sets have been connected to different phases of the supply to the welding shop, there is an increased risk of electric shock, because the differences in voltage between two welders on different phases is much greater than either to earth. Great care should be taken to avoid the situation where they can touch one another's equipment.

Dry skin offers an appreciable resistance, whereas damp or sweaty skin makes a much better connection, much increasing the hazard. Dry clothing, boots and other personal protective equipment can provide almost perfect insulation at normal welding voltages. Hence the welder working at a bench can touch live parts while standing clear of metalwork, insulated by his or her boots, without coming to any harm.

In certain situations, for example in cramped spaces such as boilers and small tanks which may be warm and damp, or in insecure positions where a shock may lead to a serious fall, extra care (such as that enforced under a permit-to-work system) is necessary. In particular, the following points require attention.

– The operator should see that the working position is dry, secure and free from dangerous obstructions, and that he or she is using adequate protective clothing.
– If damp or sweaty conditions are unavoidable, an extra check should be made to see that equipment with the lowest available open circuit voltage (OCV) is used. The welder should be kept under surveillance when he or she is handling live parts, particularly if the working area is cramped and inaccessible. A means to cut off the current must be available near the operator for use in an emergency.
– There should be adequate protection (guard rails, safety harness) if there is any risk of a fall.

Ideally, the electrical power supply should be switched off when not in use, but this is not always practicable in MMA welding so some provision should be made to accommodate the electrode holder when not in use. The practice of laying a live holder on the face screen or on a pair of gloves is not sufficient, nor is it satisfactory to suspend it by the welding cable in such a manner that it may come into contact with other equipment. An insulated hook should be provided on which it may be safely placed or a fully insulated holder should be used.

Electric heating

Where work is suspended by slings or chains, care is needed to avoid welding current flowing in them, overheating and weakening them. Overhead cranes may have their motors burnt out by excessive welding currents flowing in their windings, either directly or by inductions. It is best to ban welding on suspended work, but if it must be carried out, a separate welding return taken direct from work to set return connection and a sound earth lead attached directly to the work will minimise the problems.

Portable electric tools

Where earthed portable electric tools are in use near welding equipment, welding without a secure welding return connection to the work may fuse the earth lead of the tool, removing the protection it gives against electric shock from a breakdown in the insulation of the tool. It is better practice to use double insulated tools without an earth lead, thus avoiding the problem completely. Portable tools used in the welding shop should be checked frequently, with particular attention to earth lead and double insulation as appropriate. Earth leakage circuit breakers in addition will provide an extra measure of safety.

Electrical hazards in MMA welding

There are two types of electrode holders for manual metal arc welding:

– Type A, with complete insulation, which prevents contact between a finger and any live part of the holder. The end of an electrode in the holder is not accessible;

— Type B, with a lesser standard of insulation, but allowing the welder to choose the electrode orientation relative to the holder for welding in any position.

Type A are preferred.[151,152]

The holders must be marked with the maker's name or trade mark, the standard to which they have been made and a marking to denote the type and duty cycle. For example 'A210/60%' would indicate a type A electrode holder, rated at a current of 210 A for a 60% duty cycle. The electrodes are held in the holder by screw or spring pressure. Screw fixing is more effective and ensures a good contact but spring clamping allows electrodes to be changed much more quickly, see Fig. 11.12f and g.

Discomfort to the operator can be reduced by effective heat insulation on that part of the holder which is held in the hand. To obtain this some types employ air spaces or ducts along the handle. It is important to ensure that the cable connection is a good electrical joint and so designed that the continuing flexing of the cable will not cause wear and failure of the insulation. The jaws and connections of electrode holders should be cleaned and tightened periodically to prevent overheating. In MMA welding, the electrode holder should be isolated when changing the electrode. Otherwise, welders are relying on the insulating properties of their clothing to protect themselves.

Electrical hazards in gas-shielded welding processes

The torches, guns and other equipment for gas-shielded welding processes are more complex and more prone to accidental damage than the corresponding items for MMA welding. More care is also needed in handling the plant. A greater number of cables and hoses are necessary and these may be kept tidy by taping or strapping them together where the maker has not combined them within one outer casing. Keep the cables and hoses clear of hot plates or welds.

Gas-shielded metal arc welding

It should be noted that the consumable electrode wire is 'live' to the workpiece and earth. Therefore an insulated reel or holder must be provided and accidental contact between the wire and earthed metal work such as the frame of the set must be avoided whenever the arc power supply is likely to be switched on. In this process, accidental depression of the torch or gun switch can lead to unwanted arcing, especially where the set remains switched on when the trigger is released. Safe stowage will avoid this hazard, as well as the hazards of fire or burns from a hot gun, or of damage to the torch.

TIG welding

To strike the arc, a high frequency (HF) unit is used to provide a spark from electrode to workpiece. The workpiece terminal on the HF unit should be connected by the return leaf firmly either to the bench on which small jobs are placed or to the job itself, if this is standing on the floor. An earth lead to the bench or workpiece is essential. Earth, return and torch leads should be kept reasonably short to avoid unwanted effects such as stray sparking and to obtain satisfactory arc initiation. Cables should not be allowed to coil as this would reduce the spark.

Rubber hoses and rubber-covered cables must not be used anywhere near the HF discharge, because the ozone which it produces will rapidly rot rubber. Equipment makers can supply hoses and cables made from suitable plastics. Some plastics, such as PVC (polyvinyl chloride), conduct high frequencies to a certain extent and so rapidly break down. Rubber-covered cables are generally satisfactory for the primary connections and may be acceptable for work return and earth leads if these do not have to run near the arc.

The carbon content of some black rubber hoses for gas or water supplies can cause leakage conduction of HF currents. These and other effects can cause sparking and other unwanted behaviour, with no obvious conduction path. The HF supply cable to the torch must have special insulation to avoid stray sparking. The HF on its own will not cause an electric shock, but may startle the unwary when a spark jumps on to the hand. Dirt and metallic or other conducting dust can quickly cause a breakdown in the HF discharge unit, which should be blown out regularly to prevent such dust accumulating.

The HF spark can cause small deep burns if concentrated at a point on the skin, for example through a hole in the insulation of the torch. It is also important to feed in filler wire along the material being welded as this earths the wire and eliminates any risk of electric shock.

First aid

The first aid treatment for electric shock that causes cessation of breathing is artificial respiration, which is extremely effective if applied immediately (see Chapter 2). To avoid delay, it is desirable that all those working in areas where there is a risk of electric shock should be trained in the application of artificial respiration.

Fire risks

Sparks and spatter from the arc are always liable to ignite any flammable material in the vicinity. Care should therefore be taken to make sure that the workplace and the surrounding area are clear of anything which may catch fire. The precautions required were discussed in Chapter 3.

Noise

Gouging and deslagging are noisy operations. The operator and others nearby are liable to be exposed above the 90 dB threshold (averaged over an 8 h day). The noise level from a group of, for example, MIG welders can also exceed the 90 dB(A) exposure. If this is the case, and in any other situation where people are liable to be exposed, the levels should be assessed and hearing protection instigated (see Chapter 8).

Spatter and hot metal

The operator should make sure that his or her clothing is free from oil or grease, and that the workplace is tidy and not encumbered by any flammable material which may be ignited by sparks or spatter. When welding overhead, a cape may help to protect the welder from spatter. Protective clothing, such as aprons, gloves, spats, etc, should be inspected for burst seams or holes through which molten metal or slag may enter.

Articles which have been welded will be very hot on completion. It is recommended that these should always be clearly marked HOT in chalk or other suitable material to warn other employees who may have to handle them.

Eye injuries

Eye injuries can also occur during deslagging operations. Eye protection should be worn – for example a visor, to protect against flying
shards of slag.[99,100] If eye strain is to be avoided, an adequate standard of lighting is essential; this will also contribute to the making
of a sound weld, because the operator can see that all traces of slag
have been removed between runs.

Magnetic fields

Currents used in arc welding are frequently of the order of hundreds
of amps. This gives rise to magnetic fields, which are liable to be
larger than in most other workplaces. It is prudent for welders not
to wrap cables around themselves, both to prevent unnecessary
exposure to magnetic fields, but also to avoid being pulled off
balance. Pacemakers could be affected by large fields, but the effects
depend on the type and susceptibility of the device, and the medical
condition that it is correcting. Visitors should be alerted to the presence of fields, and any welder who is to be fitted with a pacemaker
should discuss the matter with his or her physician.

Ionising radiation

TIG welding frequently uses thoriated tungsten electrodes. The
thoria is added at 2–4% to improve the performance of the electrodes. Thorium is radioactive, giving off both alpha and gamma
radiation. Employers should consider the use of an alternative,
thorium free electrode, to reduce the risk from radioactivity.

   Bulk supplies of thoriated tungsten electrodes should be stored in
a metal cabinet or box, labelled to indicate the contents. The box
will probably supply adequate shielding against the gamma radiation. The welders should only withdraw from the store the number
of electrodes that they need each day. Regrinding of electrode tips
should be carried out on abrasive wheels with a grinding wheel suitable for hard metal. Some form of dust control should be fitted – this
may be either local exhaust ventilation or a special fluid filled grinding station. Deposits of dust should be removed daily, using a
method that does not make the dusts airborne. Spent tips and dust
from the cleaning operations should be disposed of in a sealed

container to landfill. A plastic container would be suitable, taped down to avoid dust being dispersed.

Further Reading

In the UK, there is a requirement to undertake risk assessment for work operations. In Appendix B there is a worked example for arc welding.

12

Plasma Arc Processes

Processes include plasma welding, plasma arc welding (PAW), micro-plasma welding, constricted arc welding, needle arc welding, plasma cutting, plasma spraying and plasma arc gouging.

The plasma in these processes is an ionised gas, generated by an arc between a tungsten cathode and an anode. The arc is constricted by passing it through a relatively small orifice, through which a stream of gas is also flowing. The gas may be air, argon, helium, nitrogen, hydrogen or mixtures of these gases, depending on the application. The gas is ejected as a high speed, high temperature gas jet or plasma 'flame'. This can be used for welding, sometimes making a 'keyhole' right through the weld pool, for cutting by melting metal and blowing it away, or for surfacing by entraining powder which is melted on to the surface to be treated. Both hand-held and mechanised equipment are used, Fig. 12.1.

The 'transferred' or 'direct' arc system is used, under the name of plasma arc, for most cutting, welding and gouging. In this application, the workpiece and the nozzle are both connected to the anode of the supply.

The principal use of the 'non-transferred' arc, where the nozzle is the anode, is for spraying metals and other materials, including reinforced plastics, with a variety of metallic or refractory substances such as alumina, zirconia, tungsten carbide, etc. In this instance the arc is contained wholly within the head. Metal or ceramic spraying is achieved by the introduction of wire or powder into the torch, so that it passes through the arc and is converted into a plastic or fluid state. The plastic or fluid material is then carried out of the torch by the gas stream and impinges on the component to be coated with such force that a firm bond results. The non-transferred arc is also used for gouging. A major use with either a

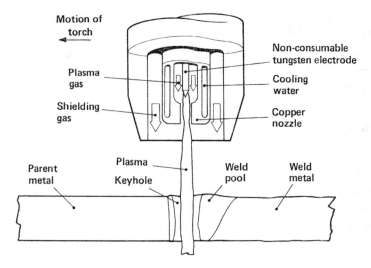

12.1   Plasma arc welding.

transferred or non-transferred arc is in welding thin materials, usually referred to as microplasma welding.

Key Hazards

Key hazards are:

– electrical
– dust, fume and gases
– noise
– radiation
– heat.

Electrical

The employer should ensure that equipment is installed by a competent electrician.

Cutting equipment requires OCVs from 100 to 400 V. Although equipment should be safe under normal operating conditions, the operator must realise the danger which is present in the use of such voltages. In addition, the technique of firing the arc by an HF (high frequency) pulse involves the danger of exposing the operator to the risk of an unpleasant shock and a painful penetrating HF burn.

Normally the supplies will be safely confined within the torch. The operator should be instructed to take particular care to maintain correct welding return and earth connections to the work, to check the torch for possible damage before switching on and never to attempt to clean the orifice by inserting a wire.

Dust, fume and gases

Spraying processes use fine powders as the consumables. The manufacturer's safety data sheet should be consulted to identify any risk to health, or any risk of explosion and fire. Many of the consumables commonly used have exposure limits, see Chapter 5.[65–68] Plasma spraying is usually carried out within a purpose-built booth which should incorporate fume extraction. It is desirable to have the fume extraction system interlocked to the spraying equipment.

In plasma cutting, the nature of the fume depends on the material being cut and the ventilation requirements should be assessed in each case. Oxides of nitrogen are also produced during cutting, particularly where nitrogen is used as the plasma gas. Ozone may be formed, particularly when cutting aluminium or stainless steel. The level of fume can be reduced by avoiding the use of excessive power and by choosing a gas that results in less fume, such as argon/hydrogen. However, in most cases, fume control measures will be needed.

The plasma cutting process tends to produce most fume below the plate, so down-draft, high-volume local exhaust ventilation may be the most effective. The design should take into consideration the possibility of damage by spatter and sparks. For hand-held plasma cutting, local exhaust ventilation of a design similar to that used for welding may be used where the assessment shows fume control to be necessary.

Modern plasma cutting installations are often equipped with a water spray covering the work area or with a shallow tank of water in which the work is submerged. The water is effective in reducing fume, noise and radiation. Cutting underwater can reduce noise levels to below 80 dB(A) and reduce fume levels to 10% of the releases when dry.

Water baths are capable of harbouring bacteria, including *legionella pneumophila*. If an assessment shows that there is a risk of legionellosis, then the system should be managed to reduce this to as low a risk as is reasonably practicable. This may include changing the water weekly and adding biocides to the tank.[153]

Water baths used when cutting aluminium will collect aluminium particles. These react with water, liberating hydrogen. The water bath should be cleaned out regularly and care should be taken to avoid areas where hydrogen can accumulate and form an explosion hazard.

Noise

The high speed at which the plasma jet leaves the torch can cause it to emit intense noise, especially in cutting and spraying activities. Noise in excess of 100 dB(A) can be produced at currents of 250 A or above. In plasma spraying processes, the purpose-built booth should be designed to attenuate the noise to a level below the action levels. Plasma cutting in water may reduce noise to 80 dB(A). However, in conditions where noise levels are above the designated action levels (see Chapter 8) effective hearing protection must be obtained and worn by people who are exposed.[105–110]

Radiation

The plasma emits both ultraviolet (UV) and visible light. Where people are liable to be exposed, eye protection will be necessary and the shade required will depend on the current and the process. Recommendations are made in both British and US standards. For cutting, full face protection is required.[98–101] Skin will also need to be protected from the UV radiation.

For microplasma work, at currents up to 15 A, a normal helmet is not necessary and a lighter type of shield, covering the face only, will be adequate, with a light filter, shade numbers 3–9 usually being chosen. Some types can accommodate magnifying lenses in addition to the filters, which may reduce eye strain in workers whose sight is less than perfect. A darker filter (higher shade number) will be required for operation in the same melting mode as normal tungsten arc gas shielded welding, at 15–100 A.

Tables 12.1 and 12.2 summarise the recommendations for filters for plasma arc work.

Heat

The non-transferred arc is contained within the torch, which consequently emits a hot gas jet even when it is not applied to the work.

Table 12.1. Filters recommended for various plasma processes[98]

Process	Current limits (A)	Suggested shade
Plasma jet cutting	To 150	11
	150–250	12
	Over 250	13
Microplasma arc welding	0.15	2.5
	0.15–0.3	3
	0.3–0.6	4
	0.6–1	5
	1–2.5	6
	2.5–5	7
	5–10	8
	10–15	9
	15–30	10
	30–60	11
	60–125	12
	125–225	13
	225–450	14
	Over 450	15

Table 12.2. Filters recommended for plasma processes[101]

Process	Current (A)	Minimum shade number	Suggested shade number for comfort
Plasma arc welding	<20	6	6–8
	20–100	8	10
	100–400	10	12
	400–800	11	14
Plasma arc cutting (only applies when the arc can be seen)	(light) <300 A	8	9
	(medium) 300–400	9	12
	(heavy) 400–800	10	14

Operators must be instructed to put down the torch in a safe place or, better still, switch off before laying it down. Apart from adequate face protection, gloves must be worn to protect the operator from sprayed particles or from molten metal during cutting. In general, exposed clothing should be non-flammable.

13

Electroslag Welding (ESW)

Electroslag welding is primarily used for producing butt welds in a vertical plane. The plates to be joined are set with a gap between them and water-cooled copper shoes or dams are placed on the sides of the joint. At the start of the joint, flux is placed at the bottom of the preparation and an arc is struck between the electrode wire or wires and a plate at the bottom. This melts the flux and the filler wire(s) are then dipped into the molten slag. The molten slag conducts electricity and heats and melts the sides of the joint and the end of the wire. As the joint fills up with metal, the dams are moved up at a rate that matches the rate of formation of the weld. Some slag is lost in the process when it forms a skin between the weld metal and the dams. This is replaced by feeding in more flux powder, Figs. 13.1 and 13.2.

Electroslag welding is therefore virtually arcless, except for the first few minutes. Even at this time the arc is largely submerged under the flux and is also shielded by the copper shoes. Thus there is no visual hazard for personnel working in the vicinity.

Key Hazards

Key hazards are:

– glare
– spatter, airborne shards of flux
– fume.

Glare

While there is no arc during the majority of the welding time, there is glare from the molten metal, and thus eye protection is required

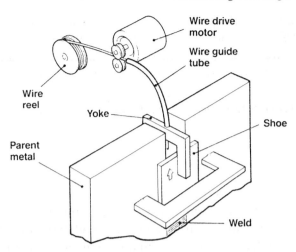

13.1   Electroslag welding.

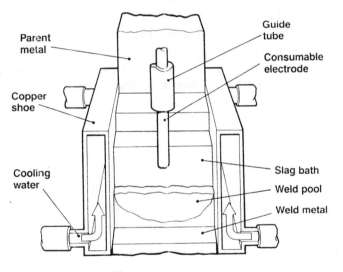

13.2   Detail of weld region.

for any person who is to approach the weld pool, perhaps to check the depth of the slag. Appropriate eyewear or face protection may be selected from the range available to welders.[98,101]

Spatter, airborne shards of slag

If the copper shoes are in correct position there is little risk of injury to the operator because no splash or spatter occurs. As a precaution against accidental spillage of flux or molten metal, caused by incorrect fitting of the shoes or by maladjustment of the welding controls, operators should wear protective aprons, gaiters and appropriate footwear. In addition, during cooling some slag splinters and fragments fly off with such force that they may injure the eyes and skin of the operator. Appropriate face protection may be selected from the normal welding or industrial ranges.[99,100]

Fume

The fluxes in common use may be of the fused, bonded or mechanically mixed type, the last being mixtures of several fused or bonded fluxes. The fused fluxes are the products of fused oxides and halide salts; the bonded fluxes consist of a mixture of finely divided oxides of manganese, aluminium, silicon, zirconium and titanium, bonded together with a suitable binder and agglomerated.

Much less fume is evolved than with most other processes, because only very small amounts of flux are actually consumed. Consequently it is safe for the process to be operated in a large, well-ventilated workshop without additional precautions, but exhaust ventilation should be provided in a small inadequately-ventilated location.

14

Resistance Welding

In resistance welding, the two pieces to be welded are pressed together by electrodes (usually copper or copper alloys) and a large electric current is passed through them. This current can be tens of thousands of amps, but the voltage could be as low as 3–4 V. The place where the two workpieces are clamped in contact has the highest resistance: hence the heating effect of the welding current is highest there and the metals melt to form a weld, see Fig. 14.1

The pressure between the electrodes may be exerted by an axial compression, or the electrodes may be rollers, where the workpiece is fed into the nip between them, see Fig. 14.2.

Once it is correctly set up, the welding cycle is fully automatic and there should be little or no spatter. However, during set up, or if the current is set too high, spatter can be produced from the weld.

There are many different designs of machine, some being fixed on a pedestal and others portable, so that they can, for instance, be mounted on a robot arm.

Flash butt welding is included in this section, since it is described in BS 499[154] as a resistance welding process. In this process two parts are brought together and a current passed across the joint until they are red hot. They are then separated and an arc is established between the ends until the metal begins to melt. The parts are then brought together under high pressures to form a joint, see Figs. 14.3 and 14.4.

Hazards

The hazards likely to be encountered in the operation of resistance welding are:

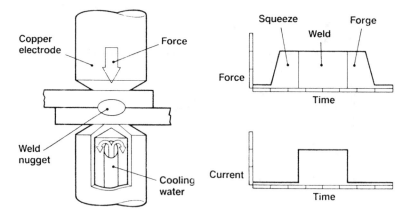

14.1   Schematic of resistance welding.

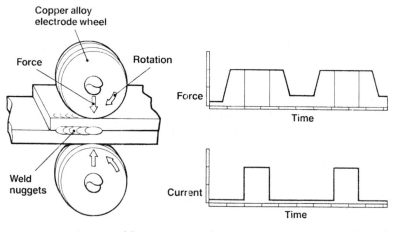

14.2   Seam welding.

-   electrical
-   magnetic field
-   hot metal
-   mechanical hazards
        cutting hazards from sheet metal,
        crushing parts of the body between the electrodes,
        being drawn into the nip point in the seam welder,
        failure of the support system for portable machines.
-   flying particles
-   fume
-   noise.

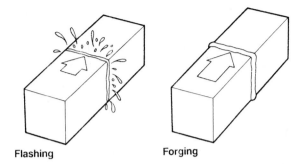

Flashing   Forging

14.3   Detail of flash butt weld region.

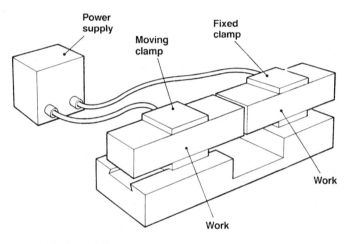

14.4   Flash welding.

A brief outline of the safety measures is given below. The equipment should be subject to periodic inspection and maintenance and the operators should be properly instructed and judged to be competent before using the equipment. Advice is given in an AWS document.[155]

Electrical hazards

The machinery must be installed according to the requirements of the electrical regulations of the locality by a competent person.[18,19,117,156] The machines often operate from a three-phase supply, which is extremely hazardous. There is a British Standard specific to the safety of this equipment.[157]

All the access panels and doors to electrical equipment which operates at hazardous voltages should be locked or interlocked, if they are accessible at the production floor, so that they cannot be opened without a key or tool. They should not be open unless the machine is being serviced and this must only be done by a competent person. Any machine that has capacitor banks at high voltages should have an interlock so that if the door is opened the bank is discharged.

External weld initiating control circuits should be at reduced voltage – not more than 120 V AC on stationary equipment or 36 V AC for portable equipment. There should be no danger from shock during the normal operation of a machine, because the voltages that are applied to the exposed and touchable current-carrying electrodes seldom exceed 20 V. The transformer secondary should be permanently grounded, or equivalent protection should be provided to protect against it becoming live as a consequence of any breakdown of insulation in the transformer.[157] Doors are sometimes provided to give access to tap changing connections and control cabinets. Such doors should be fitted with safety interlocks, if these allow access to potentially hazardous voltages.

There is one process, HF resistance welding, in which there are high voltages on exposed welding electrodes. The frequency of the electric current is so high, usually about 400 kHz, that there is little danger of electric shock if the terminals are touched. However, small but very deep and painful burns are likely to occur where the current enters and leaves the body and for this reason adequate precautions should be taken to prevent contact with these terminals while they are live.

Magnetic fields

The high currents used in resistance welding give rise to correspondingly high magnetic fields, particularly in large machines of the DC or three-phase (low-frequency output) types.

Static magnetic fields are subject to advisory exposure limits (see Chapter 1). When the current is pulsed, as for repetitive spot welding or for seam welding, the changing magnetic fields could affect the operation of some body implants, including some electronic heart pacemakers. Changing fields induce small voltages in the pacemaker unit, which could indicate normal heart function to the pacemaker, inhibiting pacemaker pulses. The patient's heart might slow down or stop, causing him or her to faint. Fortunately, normal functioning should be rapidly recovered on removal of the field. Normal walls or partitions will not screen off the low frequency magnetic fields most likely to cause adverse effects, but the fields fall off to negligible values at 10 m. Similar but smaller effects may arise from arc welding, with a corresponding distance of 3 m. A suggested plan of action is:

- Advise known pacemaker users to keep clear of the arc and resistance welding if possible.
- Include advice in induction procedure for new employees.
- Place notices to warn visitors of the risk.

Since adverse effects are rare in practice, if a pacemaker user wishes to work near or with welding equipment they should ask their general practitioner or hospital consultant for advice; they may be able to arrange a trial run under medical supervision.

Hot metal

Sometimes no visible heat is produced during the welding operation, consequently there is a tendency to think the welded parts are not hot. Burns will result from careless handling of hot assemblies.

Mechanical hazards

Resistance welding is extensively used for guillotined, blanked or pressed components which may have very sharp edges or burrs. The incidence of cuts and lacerations can be minimised by deburring the components before they are taken to the welding position and by wearing protective gloves or gauntlets.

Resistance welding machines invariably have at least one electrode which moves with considerable force; in this respect they closely resemble a power press. If a machine is operated while a finger or hand is between the electrodes or platens, severe crushing will result. The risk of crushing should be considered for each machine and a suitable means must be devised to safeguard the operator. Seam welding machines require similar protection against nipping hazards between the wheel electrodes. Unless the workpieces are of such configuration that the hands are remote from the point of operation, then the safeguards should be one of the following:

- guards or fixtures to prevent the hand entering the danger zone
- two-handed controls
- latches
- presence sensing devices
- or any similar measure.

A recent development, applicable to a wide range of machines and with little disturbance to normal operation, uses a modified air pressure system. The initial approach is with a low pressure, which cannot produce sufficient force to cause significant injury to a trapped hand or finger. When the electrode gap has been reduced to 6–8 mm, in the absence of any obstruction, the full air pressure is applied to produce the normal weld squeeze force. An alternative is a probe near or surrounding the electrode which switches off the machine if it encounters an obstruction.

For those machines that have a sequence that takes 3 s or more to complete, and that have mechanical movements that could be hazardous if the guards were removed, emergency stop buttons should be provided within reach of the operator and anyone who may get trapped, unless the buttons themselves would create additional hazards to persons.

Some production jobs on assemblies of a complex shape, such as used in the construction of motor vehicles, may be carried out by robots; the hazards presented are discussed in Chapter 9.

Flying particles

Particles of hot or molten metal will not fly out of spot, seam or projection welds if the material and welding conditions are ideal, but such particles are frequently expelled in production work and the greatest danger is to the eyes of the operator or a passer-by. Loose metal parts should not be left in the throat of the machine because they are liable to be projected from it with some velocity by electromagnetic forces.

The flash welding process inherently produces a considerable quantity of flying red hot particles which may travel up to 6 m and which may enter unprotected eyes and exposed skin with some force. Fire-resistant screens should be used to protect persons in the vicinity and precautions should be taken against an outbreak of fire. Whenever there is a risk from flying particles people entering the area should wear goggles and protective clothing.

Fume

If the articles to be resistance welded are free from dirt, oil and other extraneous material, little fume should arise during the operation. The amount of fume is unlikely to be great and will disperse quickly in a well-ventilated shop, but in small and inadequately-ventilated workplaces it may be necessary to provide local exhaust ventilation. Paint will usually prevent the welding current flowing and so will be cleaned off the work in the weld region.

It will be necessary to ascertain the nature of any fume to which workers may be exposed, and establish that it does not exceed the limits laid down (Chapter 5). The use of cadmium is now restricted, but if this material is present on the parts to be welded extra care should be taken in controlling exposure to the fume since it is extremely toxic.

Noise

Certain resistance welding machines, for instance some that are used for flash welding, produce excessive noise which may be harmful to hearing. The noise that arises from the machine, such as compressed air exhausts, actuator cylinders or loose transformer laminations should, wherever it is reasonably practicable, be reduced at source. If this fails to reduce the noise exposure to employees enough to comply with the standards, then ear protection will be required, see Chapter 8.

15

Thermit Welding

Thermit, as used for welding applications, is a mixture composed of finely divided aluminium and iron oxide. Alloying elements such as manganese, carbon, molybdenum, nickel, vanadium, chromium and titanium can be added as required to give the strength and hardness required of the finished weld.

A mould is fabricated around the parts to be welded. A crucible is placed above the mould, containing the thermit mixture. When the mixture in the crucible is ignited, the aluminium reacts with the iron oxide to form aluminium oxide and iron. The reaction gives out a large quantity of heat, which ensures that all the material reacts and the iron is melted and its temperature raised to around 2500–3000 °C.

After the chemical reaction is complete, time is allowed for the dense steel and lighter slag to separate and the crucible is tapped 20–60 s after ignition. The molten steel flows into the mould and fusion takes place with the parts to be welded, Fig. 15.1.

Since the ordinary thermit charge has a high ignition point, it is necessary to place a low ignition point powder on top of the charge. The ignition powder in general use is composed of fine aluminium powder with a peroxide, chlorate or chromate. For most welds the end of the pieces to be welded will be preheated to a temperature of up to 1000 °C. Preheating torches, which are inserted into the heating gate of the mould, are fired by propane or butane and air, or by propane and oxygen. The process can be operated in the open air or under cover. It is frequently used for welding lengths of railway line.

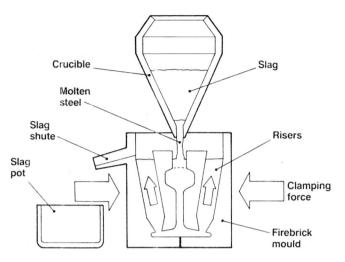

15.1   Thermit welding.

Key Hazards

Key hazards are:

— violent reaction and ejection of the thermit mixture if it is damp
— fire or explosion from the process or the dusts
— escape of molten metal due to failure of the mould/support structures
— fume
— infrared radiation, spatter and sparks.

Thermit mixture

The thermit mixture must be kept dry at all times. If it has become damp in storage and even if it has subsequently dried out, the aluminium–oxygen balance of the mixture will have been upset; this will not only affect the quality of the thermit steel and consequently the strength of the weld, but can cause the violent evolution of hydrogen or carbon dioxide gas. If the mixture is wet when the reaction starts, steam will be generated inside the crucible; either steam or gas will cause molten slag or steel to be ejected.

Fire and explosion hazards

Care must be taken in choosing a welding site to ensure that fire risks, such as structures made from flammable materials, are sufficiently far from the equipment to be unaffected by splashes, or radiant heat from the crucible or mould.

Work should preferably be carried out under cover to protect the equipment and the job from the weather. Buildings that have wooden floors or wood in the roof structure should be avoided. The building chosen should be dry and well ventilated. If an existing building is used it should be of sufficient height to allow the equipment to be set up well clear of any timber structures. If the completed job requires the use of lifting tackle, space must be available for this to be operated. Space must also be available around the job to allow free and unobstructed movement at all times for the employees engaged there.

Thermit mixtures should be stored separate from the ignition powders or igniters used for starting the thermit reaction. Igniters should, if possible, be stored in a separate building. The ignition temperature of the igniter or ignition powder is about 200–300 °C and of the thermit mixture about 1000 °C.

Thermit for repair welding is normally supplied in linen bags, containing 10 kg of mixture, which are packed in steel drums. The steel drums are lined with polythene and the thermit powder should be stored in them until it is required for use. For rail welding, preweighed portions are supplied in polythene bags. Thermit should always be stored away from materials of flammable nature that may be liable to self-ignition. Some thermit mixtures and aluminium powders have been found to present a dust explosion hazard. Because of this, workrooms must be kept clean and accumulations of dust avoided, paying particular attention to any areas not readily visible.

The mould

For welding of a repetitive nature where the pieces to be welded are of small dimensions a preformed refractory mould may be used. For non-repetitive work it is usual for a steel case to be built around the job, which is rammed with a suitable grade of sand to form the mould. It is essential for the mould box to be of sufficient strength to withstand both preheating and the added weight of thermit steel at the time the weld is made. If not strong enough, any movement or buckling of the mould box may result in a cracked mould which will allow the molten steel to escape, at a temperature of about 1800 °C. The moulding sand must be moistened to render it workable before mould making. The moulding sand will dry out during the preheating operation if it has the correct permeability factor and no trouble resulting from steam evolution should be encountered.

The 'lost wax' process is generally used in repair welding. The wax melting out produces fumes which will need good general or local ventilation to remove.

The crucible support

It is usual for the crucible support to be built in the form of a bridge spanning the mould box. Tubular steel scaffolding is often used for this purpose for reasons of economy and for ease of transport, storage, etc. The structure must be of sufficient strength to withstand the weight of the crucible and thermit charge. It will be subjected to vibration caused by turbulence in the crucible during the thermit reaction and to heat arising from waste gas issuing from vents in the mould during the preheating operation. These factors must be taken into consideration during construction. The structure can be shielded to some extent by the use of suitable deflector plates above the waste gas vents in the mould. All joints in the structure, particularly if made of tubular scaffolding, must be examined at intervals during the preheating operation and tightened if necessary.

Ventilation

The gas-burning equipment used for preheating may produce a toxic hazard, especially if the equipment is incorrectly used and combustion of the fuel is incomplete. In this connection the following possibilities must be avoided: leaky gas connections and the emission of waste gas or paraffin fumes. Generally, an adequately ventilated workshop should be sufficient to avoid discomfort or hazard to the operators, but if this is not the case, forced ventilation will be necessary during the preheating and reaction stages.

Where thermit welding is used to join austenitic manganese steel the powder contains manganese to provide a matching weld metal (12–14%) and substantial toxic manganese fume is evolved during the reaction. Manganese fume has strict exposure limits: $1\,\mathrm{mg\,m^{-3}}$ in the UK for an 8h day, $5\,\mathrm{mg\,m^{-3}}$ in the USA (ceiling value), measured in the breathing zone.[65,66] It is essential to restrict exposure to the lowest level reasonably practicable, below the statutory limits.

Protective equipment

In rail welding the process equipment is relatively close to the ground and it is necessary for the welder to look into the crucible before tapping. The welder must therefore wear suitable eye protection. The main radiation hazard is infrared emitted by the molten steel. Adequate protection is given by a filter to BS EN 169 shade 4 to 6[98] (or the USA equivalent[101]). A young person with good visual acuity can use a darker filter (5 or 6) which provides an improved view; but older people may prefer a shade 4. The welder will also need protection against splashes of molten metal or slag and this may be gained by wearing a face shield of a pattern appropriate for molten metal service over the welding goggles.[99,100]

In repair welding the equipment is generally larger and assembled at greater distances from the floor. The molten metal pouring from the crucible is at about the welder's eye level and similar filters in a helmet or hand shield are recommended. Protective clothing such as aprons, hoods, leg shields and gauntlets should be provided and used as appropriate for the job. The welder must use a clear face shield when trimming the weld as a protection against flying hot particles and sand.

16

Electron Beam Welding

Electron beam welding directs a focused beam of electrons onto the workpiece, where the electrons lose their kinetic energy to the metal, Fig. 16.1. The electrons may be accelerated through very high voltages, up to 200 kV, in order to give them enough energy to heat the workpiece sufficiently. As a by-product of the process, X-rays are generated by the impact of the electron beam on the workpiece.

The essential parts of an electron beam welding machine are an electron gun, a means of focusing and accelerating the beam, the vacuum system (if used) and a traverse for the workpiece. Since these machines are potentially very hazardous, modifications must be done only by the manufacturer or by a qualified service technician.

Before installing an electron beam welder for the first time in the United Kingdom, the Health and Safety Executive must be notified at least 28 days in advance. The employer will need to undertake a risk assessment and formulate a set of rules for the operation of the equipment to protect their employees against exposure to radiation, both during normal use and as a result of any foreseeable accident. The employer will need to appoint a radiation protection advisor and one or more radiation protection supervisors to assist the employer in meeting the legal requirement.[103]

The US reader should check the local regulatory requirements before installing and commissioning such equipment and read the guidance relating to its use.[104,158–160]

Three types of machine are available: high vacuum, medium vacuum and non-vacuum.

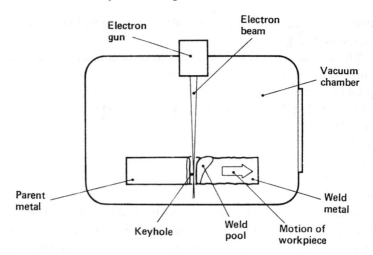

16.1   Electron beam welding.

Key Hazards

Key hazards are:

- electrical
- radiation
- fume or dusts
- physical hazards.

Electrical hazards

The safety of the equipment on the exterior of the unit is largely the responsibility of the designer and the manufacturer. To protect against electric shock the user should be satisfied that a system of interlocks has been fitted to the various cabinets and that these inter-locks cannot be jammed in an unsafe position. The cabinets should be marked with symbols indicating the electrical hazard to warn against opening them. A good and reliable system of earthing is essential, linking together the working chamber, electrical supply and control cabinets, and the pumping units (if any).

Simple types of electron gun employ the workpiece as the anode. More refined guns have an anode constructed as part of the electron gun, the workpiece being at the same potential as the anode. For safety in operation the anode of the system is normally at earth potential. Connections made in this way ensure that the workpiece is at earth potential and that the only live parts of the high tension (HT) supply are the cathode, any subsidiary anodes in a stepped potential below earth and the connecting cables inside the vacuum chamber.

The whole of the HT circuit should be in earthed metal enclosures and HT connecting cables should be of a suitable type, such as pliable armoured cable with proper terminals (armour clamps, glands, etc). Such a design is adopted in most cases to ensure a reliable HT supply, free from the risk of insulation breakdown. It also ensures maximum safety for the operator when handling anything external to the welding chamber.

Ensure that the equipment is completely switched off and isolated before the operator enters the chamber for the purposes of cleaning, adjusting the electron gun, or loading and unloading the assembly to be welded. A grounding rod should be used to discharge any static charge on the gun before working on it and to prevent any build up during work. At all times when the chamber is at atmospheric pressure, the HT supply should be isolated from the welding chamber. An effective means of securing isolation is to ensure that the vacuum chamber is fitted with a capsule-type vacuum switch wired into a contactor which controls the supply at HT to the electron gun. In this way the HT supply is automatically isolated when air is let into the chamber.

With high volume production equipment it is not feasible to switch off the high voltage supply: instead it is 'ramped' up and down over a period of seconds; also a vacuum valve isolates the gun column from the chamber. As an additional safeguard a door switch should be fitted to the vacuum chamber wired so as to disconnect the HT supply.

The method of interlocking the doors with the electrical circuits must receive very careful consideration. It is suggested that the standard relating to control systems is consulted.[161] Most modern equipment has the HT supply enclosed within an oil tank inside the cubicle, so there is no risk from this, but normal mains voltage hazards may still exist.

In any equipment large enough for a person to enter there is always the danger of a person being trapped inside or of powerful work handling gear or door motors being energised by someone who cannot see the person inside. Protection is given if those entering the equipment carry an interlock key without which the equipment cannot be switched on (such as in the Castell key system). This can be applied both to cabinets containing electrical gear and/or to the vacuum chamber as appropriate. There is also justification for fitting a safety fence in some circumstances.

Radiation hazards

An unseen hazard associated with the use of electron beams at high voltages is the generation of X-rays. It is imperative that sufficient shielding is provided to protect the operator from the possibility of X-rays penetrating the walls and windows of the vacuum chamber. X-radiation is more penetrating at shorter wavelengths, which correspond to the higher voltages. It is important when specifying the shielding to take into account the maximum radiation that the machine can produce and establish a level of shielding that will be adequate. The intensity of the radiation is proportional to the square of the voltage and to the beam current.

Exposure to radiation must be reduced to the lowest level reasonably practicable. Non-vacuum equipment should have automatic warning devices to warn that an electron beam is about to be generated. In general, for voltages less than 50 kV, the wall thickness of a vacuum chamber determined by structural considerations will provide sufficient inherent shielding. For example for a 30 kV machine, the dose rate outside the chamber was less than $0.5\,\mu Sv\,h^{-1}$ above background. As the voltage is increased, particularly above 100 kV, the shielding offered by the walls of the vessel must be supplemented, usually by the addition of a suitable thickness of lead. Initial radiation measurement and control will normally be the equipment manufacturer's responsibility. However, at the time of installation, the employer should do his own check of the electron beam welding machine.

The manufacturer of the equipment should test it after installation and provide a record of the tests, along with whatever information is needed to the user to ensure continued safe operation. Any parts of the machine providing shielding against ionising radiation should be interlocked in such a way that, unless they are in position, the machine cannot be energised. Should such an interlock not be practicable, a system of work must be devised such that any shields removed for maintenance, repair or other purposes are replaced before normal operations are resumed. This could be implemented by a permit-to-work system, with a check sheet.

Fume and dusts

During evacuation of the chamber the vacuum pump must remove a volume of air equivalent to the volume of the chamber. As the mechanical vacuum pump which performs this duty operates under oil for sealing and lubrication, the oil is agitated and the escaping air carries a fine mist of oil. This is unpleasant to breathe in, deposits an oil film on surfaces near the exhaust and carries a possible toxic hazard from the major oily constituents and any minor additives. Control of this oil mist is generally desirable and may be achieved by filtration and/or exhausting the air outside the buildings, clear of any windows. Filters will need to be selected which will pass the large volumes of air involved without creating excessive back pressure on the pump. Special purpose oil mist separators are commercially available. The vacuum pump exhaust will carry any toxic material present in the chamber atmosphere, so it must be suitably sited and/or filtered to avoid hazard.

When welding a metal, evaporation will take place to an amount dependent upon the vapour pressure of its constituent elements. Normally the vapour will condense to form films on the interior of the vacuum chamber, or fine dusts in the chamber. It is important to assess the properties of the metal dusts that are formed, to discover whether they may be toxic, flammable or explosive. This will affect the procedures that will be drawn up for entering the chamber after welding and for the cleaning regime.

If the materials being welded are toxic, for example beryllium, plutonium, etc, great care must be taken when opening the chamber to protect the operator from any dust cloud which may have arisen during the admission of air. To secure this protection it is recommended that an exhaust system is fitted to the chamber via a vacuum valve. Thus the exhaust system is automatically opened to an extractor immediately the chamber door is opened. At the same time clean air is drawn into the chamber and dusts are removed in a direction away from the operator. Care will be needed in the handling and disposal of the filters. Exhaust ventilation used for the removal of fine metal dust from the chamber may well constitute a dust explosion hazard for certain metals. If this is the case, the collectors must be sited with care, and explosion relief panels should be included.

The potential exposure of the operators to hazardous substances must be addressed. If the engineering control measures detailed above are not sufficient to restrict exposure to levels below the exposure limits, then respiratory protective equipment will be necessary. When cleaning out the inside surfaces of the chamber, the potential for exposure to the disturbed dust film must also be assessed and controlled.

Physical hazards

The cooling rate of a material in a vacuum is slower than if it were in air, where it loses heat not only by radiation but also by convection and conduction. Thus, if the structure being welded is not able to cool by direct conduction to its supporting jig, adequate cooling time should be allowed before handling. Signs should be positioned to warn others of the hazard.

Chambers in which electron beam welding is carried out are frequently very large and consequently persons can enter them. It is essential to arrange a system of work, such as a Castell key system for entering the chamber, to ensure that no person is inside the chamber when the machinery is started up. In addition the

16.2   Anyone trapped inside the Welding Institute's $150\,m^3$ electron beam welding chamber can cut off the pumps by pulling a safety lanyard. The chamber walls are covered on the outside with lead sheet for X-ray shielding.

chamber should have a prominent panic wire that is interlocked to the vacuum pumps (if any) to shut them down and vent the system to air, and to cut off the HT circuits of the electron beam, see Fig. 16.2.

17

Friction Welding

Friction welding is a non-fusion process. One component is rubbed against the other under an axial force; friction at the interface develops heat, raising the temperature. At an appropriate point, rubbing is stopped and the axial force forges the joint, Fig. 17.1.

Originally this method of joining was only applied to round parts, such as shafts, but now machines are available with a reciprocating action that can weld items that are other shapes. There are also machines that rotate a hardfacing consumable against a substrate for surfacing, and machines that 'stir' two parent materials that are butted together to form a joint between them. Joining takes place below the melting temperature of the parts, and fluxes, fillers and protective atmospheres are not required.

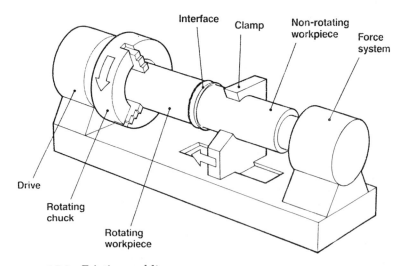

17.1   Friction welding.

The chief hazards associated with friction welding are mechanical. In order to generate the friction to heat the materials motion must be rapid. The forging forces to create the joint are also large. Many of the prototype machines were converted milling machines.

Machinery must be designed and operated so that access to dangerous parts, such as rotating shafts is not possible.[7,162] Practical advice on the design of guards is given in a number of publications.[163,164] Guards should not be easily defeated and should be interlocked, so that if they are opened the machine shuts down. Therefore if the operator needs to gain access to the workpiece, for instance to measure the speed of rotation of the shaft during welding, it should be arranged that this is done with the guard closed.

18

Laser Welding and Cutting

Lasers produce parallel, generally monochromatic, beams of non-ionising radiation, which may be visible, ultraviolet or infrared. This radiation can be focused to a point to produce intense heating for laser welding and cutting. Lasers can yield high production rates and can make deep penetration welds, with a minimal heat affected zone, including joints between dissimilar metals, see Fig. 18.1.

There are several types of laser that may be encountered in welding and cutting, identified by their active medium. Table 18.1 illustrates some typical uses.

Key Hazards

Key hazards are:

- laser radiation
- electrical
- fume and gas
- fire
- mechanical hazards from the workpiece handling system (or a fibre optic beam delivery).

Laser radiation

The damage that laser radiation can do depends on its wavelength and its power. The principle areas of concern are the effects on the eye and the skin. Ultraviolet light affects primarily the cornea (surface of the eye) causing arc eye, while visible light and near infrared (roughly 400 nm to 1.4 µm) can enter the eye and will be incident on the retina. Beyond 1.4 µm, the cornea is, once more, the affected part.

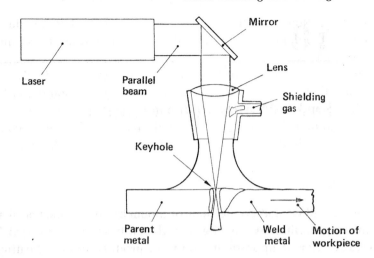

18.1   Laser welding.

Table 18.1. Typical uses for lasers

Laser type	Output wavelength	Typical applications
Helium–neon	633 nm (red)	Alignment
Ruby	694 nm (red)	(Now relatively uncommon) hole drilling
Neodymium/YAG	1.06 µm (near infrared)	Welding and cutting Drilling Can be used with a fibre-optic delivery and robot manipulation
Carbon dioxide	10.6 µm (mid infrared)	Welding Cutting of metals, ceramics, polymers and cloth Heat treatment Etching
Excimer lasers	Ultraviolet	Precision machining, drilling, polymer ablation

Ultraviolet light will cause sunburn and premature ageing of the skin. Laser light in the visible region may cause thermal damage and changes in skin pigmentation; in the infrared the skin will be burnt. The damage that the laser radiation can do is dependent on the power of the laser. The wavelengths that can reach the retina are particularly hazardous, since the eye focuses the radiation to a small spot. This can result in almost instantaneous irreversible damage to the sight. Note that the results of exposure to UV are delayed, whereas the damage done by visible and infrared lasers is immediately apparent.

Because the hazards of lasers are not always immediately apparent and they can be so severe, both the American Standards[165] and the UK Standards[166,167] require the employer to appoint a laser safety officer, to monitor the use of lasers, design the engineering controls that are required and ensure that safe practices are adopted.

Laser classification

Lasers are classified on the basis of the maximum output power and wavelength, Table 18.2. The manufacturer should label products accordingly with appropriate warning labels. The labels are black on yellow.

Class 1 is safe under all reasonably foreseeable circumstances, classes 2 and 3 are more hazardous, and class 4 is the most hazardous. In the latter class, the beam is hazardous even when dif-

Table 18.2. Classification of laser products

Class	Description	Type and content of labels
1	Inherently safe Also a higher class laser with engineering controls such that no radiation greater than class 1 can be accessed	'Class 1 laser product' (optional)
2	Lasers emitting visible radiation, where the human eye is protected from damage by the aversion response	Laser radiation. Do not stare into beam. Class 2 laser product
3A*	Lasers that are safe for viewing with the unaided eye. The use of optical aids (e.g. microscopes) may be hazardous	Laser radiation. Do not stare into beam or view directly with optical instruments. Class 3A product
3B	Direct intrabeam viewing is hazardous. Diffuse reflections are normally safe	Laser radiation. Avoid exposure to beam. Class 3B laser product. Aperture to be labelled
4	Lasers that can produce hazardous diffuse reflections. They may cause skin injuries and constitute a fire hazard	Laser radiation. Avoid eye or skin exposure to direct or scattered radiation. Class 4 laser product. Aperture to be labelled

* Note that a class 3A laser under the US classification is not identical to the UK classification.

fusely reflected from a surface and also represents a significant fire hazard.

Helium-neon lasers have too low a power output to be of any use directly for welding and cutting but may be used in conjunction with other types to align optical components and workpieces. Since the output is in the visible spectrum, beam paths and point of focus are easily seen with the naked eye. They will normally comply with class 1 or 2. The majority of the materials processing lasers are in class 4 and powers can be up to tens of kilowatts.

Precautions for class 4 lasers

The following facilities should be provided. Many will normally be built-in by the manufacturer of the equipment and described and illustrated in its accompanying documentation. However, the incorporation of the laser into a working cell will normally be the responsibility of the employer, who will also be responsible for implementing the procedures.

An emission indicator device must clearly show when the laser is energised. The laser should only be operated in a controlled area. The laser radiation should be confined within a protective enclosure where reasonably practicable. Warning signs of the approved type should be placed at the points of access to the work area. Signs will warn of the danger and the class of the laser and will include the standard hazard symbol, Fig. 18.2.

There should be a remote interlock connected to room, door or fixture circuits, to cut off power if doors are left open or are opened while the laser is energised. In the case of a pulsed laser, these interlocks should also cause the stored energy to be discharged. The laser should be operated by remote control if possible so that personnel are clear of the work area. Operation should only be possible by the

18.2   Hazard warning symbol for laser radiation (black on yellow background).

use of a key and this key should be removed when the equipment is not in use to prevent unauthorised operation.

As far as possible, beams should be enclosed, but if unprotected, beams should be well above or below eye level. Specular (mirror-like) reflections should be avoided by mounting mirrors and lenses rigidly and by avoiding polished surfaces, such as personal jewellery and tools, in the vicinity of the beam path. Note that surfaces which appear quite rough in visible light may give specular reflections at a wavelength of 10.6 µm. For class 4 lasers, even diffuse reflections have the potential to be harmful and should be avoided. Alignment of the beam should be done using low power lasers.

A beam attenuator and/or beam stop, capable of safely absorbing the available energy should be placed to cut down the energy as much as practicable and to terminate any residual beams. Panels which are removable for servicing, etc, must have interlocks to prevent exposure if they are removed. If any interlock is provided with an override mechanism, there must be a warning notice and a visible or audible warning.

Personnel who may be exposed to radiation should wear clothing with appropriate resistance to fire. All operation and maintenance personnel should receive appropriate training including:

– operating procedures
– hazard control procedures
– personnel protection
– accident reporting
– bio-effects of the laser.

Protective filters

Where engineering and administrative controls cannot substantially eliminate the risk of exposure in excess of the maximum permissible exposure, eye protection should be provided, clearly marked to ensure proper choice.[165,168,169] However, it should be noted that eye protection is the last resort – engineering controls to prevent exposure should always take precedence. Eye protection is unlikely to be able to protect the wearer against exposure to the direct beam.

The windows to the engineering enclosure of the beam may be furnished with filter material, to allow a view of the workpiece while ensuring that laser radiation cannot be transmitted. As with arc welding, good room illumination and matt, light coloured walls are desirable to give operators a good view through protective filters. As only special materials will transmit the 10.6 μm radiation of the carbon dioxide laser, there is little difficulty in making some or all of the protective enclosure of plastic sheet transparent to visible light. Note however, that if the full beam impinges on the screen material it may melt through. Operators should be further protected by wearing clear plastic visors.

Electrical hazards

All equipment should be installed in accordance with the National Codes.[18,19,156] Pulsed ruby and neodymium lasers draw their electrical input energy from a capacitor bank or pulse-forming network charged to several kilovolts. Interlocks are required to prevent live parts being touched until the mains supply has been cut off and the stored energy safely dissipated.

Carbon dioxide lasers have a substantial high voltage supply, often enclosed within the case needed to protect the optical components. If a flexible cable carries the supply from a separate power unit, it must be suitably protected and inspected regularly for damage.

Water leaks must always be repaired promptly, especially when the water is routed with the electrical cables.

Only those suitably trained and competent should attempt to work on the power supply of a laser. Fatal accidents have occurred.

Power supplies should be provided with an automatic discharge and grounding circuit that is actuated when the laser is turned off. There should be discharge and grounding interlocks on all access panels. The maintenance engineer should be provided with a grounding rod for manual verification of complete discharge. He or she should wear safety glasses when undertaking this operation since explosion-like discharges can take place. There should be grounding straps to prevent charge from building up.

Fume and gas

Substantial amounts of fume may be emitted in work with high power carbon dioxide lasers, particularly when cutting with the use of oxygen. Fume hazards and preventive action are discussed in Chapters 5 and 6; the enclosure that protects the operators from the laser beam should incorporate an appropriate extractor as necessary.

Gases in cylinders present the hazards described in Chapter 4, and require safe storage and handling.

Fire

High power class 4 beams can set articles on fire. Ensure that the beam is always stopped at the end of its useful path by a beam stop that absorbs the radiation safely without catching fire. Ensure a good standard of housekeeping so that there is no combustible rubbish which the beam could set on fire.

Mechanical hazards

Lasers are capable of very high welding speeds and therefore it follows that either the beam must be moved rapidly, or the work-piece. For gas lasers, it is common to use an X-Y traverse to manipulate the workpiece. Fibre optic delivery of the Nd-YAG laser allows the laser itself to be manipulated by a programmable robot.

When working at full speed, the manipulators should be within the laser enclosure, and therefore separated from the work force. However, it is usually necessary to program the motion in advance. This will normally be done at a reduced speed, with such devices as 'hold to run' controls. Chapter 9 and the references relating to robots give further information.[120–122]

19

Brazing and Braze Welding

Brazing, which includes hard soldering and silver soldering, involves joining two closely fitting surfaces by a filler metal which has a lower melting point than that of the parent metal. The filler metal is drawn into the joint by capillary action. Braze welding is similar; the filler is a lower melting point alloy than the parent metal, but in this process it does not enter the joint by capillary action. There is a conventional cut-off at 450 °C, below which the processes are known as soldering. These are dealt with in Chapter 20.

Brazing Processes

Five basic heating methods are used in brazing:

- torch
- induction
- resistance
- furnace
- salt and flux bath.

Each is accompanied by different hazards, which are dealt with under the process description. The hazards that are common to all are fume and chemical hazards from the fluxes and the hazards associated with the preparation of parts to be brazed, which are dealt with after the processes are described.

Torch brazing or braze welding

In this method the filler metal is either preplaced in the joint as an insert or externally applied from a rod and a flux is used to shield the brazing operation from the atmosphere. A heating flame is produced by a torch supplied with an oxygen-fuel gas or air-fuel gas

169

mixture, and the filler metal is fused by heat conducted from the hot component parts. Gas mixtures often used are air-natural gas, oxygen-natural gas, oxypropane and oxyacetylene, the oxyacetylene flame having the highest temperature. These gases should be stored as described in Chapter 4. The hoses and accessories should be as described in Chapter 10.

In torch brazing the operators tend to bend over the job to have a closer view of the work and hence are liable to inhale any of the fumes emitted from the operation. An uncomfortable glare will be experienced during continual torch brazing. This can be overcome by the use of goggles, fitted with appropriate filters; the types with an 'a' suffix are made to minimise transmission of the glare produced by fluxes. Shade 3 or 4 should be sufficient. Filters and shades are discussed in Chapter 10, see also Tables 10.2 and 10.3.

Mechanised torch brazing employs a number of fixed torches and the work is moved through or into the heating flames. In this method the operator assembles the components away from the brazing operation and is therefore less exposed to the fumes which may be produced.

Braze welding is a technique that is similar in some ways to both welding and brazing. It is mostly used to join iron and steel components, but it can be applied satisfactorily to certain copper alloys.

Basically the process consists of heating the assembly with the cone of a gas flame as in welding, but using a lower melting point filler alloy, as in brazing, to provide the connection between the two parts. The filler metal is continuously fed into a pool of molten filler metal using a technique similar to gas welding. Bonding of the joint results from wetting the unmelted surfaces and interdiffusion of the filler and parent metals. Many combinations of gases may be employed to produce the heating flame but the three most common are oxyacetylene, oxyhydrogen and oxypropane. The process is such that dissimilar metals may be joined, provided the fusion temperature of both parent metals is above 950 °C.

Many filler alloys are used but they are commonly rich in copper with small additions of nickel, silicon, tin, manganese, iron, aluminium and lead. The application of flux which is likely to be of the borax type is necessary. The filler alloys are fed into an extremely hot flame and considerable fume may be evolved which should be controlled if necessary by the provision of local exhaust ventilation, as copper fume is a potent cause of metal fume fever (see Chapters 5 and 6).

Table 19.1. Filter shades for braze welding[98]

Work	≤70 litres acetylene per hour	Between 70 and 200 l/h^{-1}	Between 200 and 800 l/h^{-1}	Over 800 l/h^{-1}
Braze welding of steels, copper etc	4	5	6	7

The operator can be continually looking at the hot cone of the flame and incorrect use of the blowpipe may result in the filler metal splashing; protective clothing and goggles should be provided. Table 19.1 shows the shades recommended.

Induction brazing

In this method heat is produced by eddy currents induced in the components from an HF current passing through a water-cooled coil encircling the joint; heating is intense and localised. A paste or powder flux is commonly used, the filler metal is preplaced in the joint and the components are assembled for brazing in a jig. The operator loads one jig while another assembly is being brazed, and is unlikely to be in close contact with any fumes produced, see Fig. 19.1.

Most of the electrical equipment is likely to be housed in a cabinet which should not be opened until the power has been switched off. With high frequency (HF) equipment the operator will not receive a harmful electric shock from touching the work coil, but may suffer

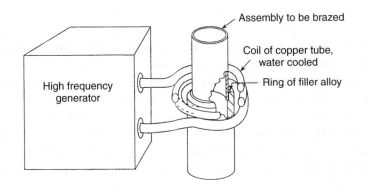

19.1   Induction brazing.

an HF burn which will be deep, painful and slow to heal. Precautions should therefore be taken to avoid contact with, or close proximity to, the work coil. Operators should not wear rings if they are able to bring their hands near to the work coil while is it energised. Rings will be heated just like the work.

Resistance brazing

Brazing by resistance heating is employed for the rapid assembly of small components. One method makes use of a resistance welding or heating machine equipped with a special copper alloy electrode. Heat for brazing is generated by the resistance offered to the passage of a heavy electrical current through the joint. Heating is extremely rapid and very little fume is evolved, as the joint is enclosed and under pressure. If the equipment is incorrectly set, sparks may be emitted during the passage of the current.

The other method employs two carbon electrodes which make contact with either side of the joint area. Passage of an electric current causes the electrodes to glow, thus heating the joint by conduction. Flux absorbed by the carbon electrodes may cause some fuming, but in most cases the operator does not have to bend over the work.

Resistance heating uses low secondary voltages so that there is little risk of electric shock (see also Chapter 14). There is a risk of spatter, which may be overcome by wearing personal protective clothing.

Furnace brazing

The components are assembled with brazing alloy and heated in a furnace of the muffle type, which may have an inert or reducing gas atmosphere. The furnace may be heated by electricity, gas or other fuel. While a conventional flux is sometimes used, a controlled atmosphere is normally employed. This atmosphere will often contain asphyxiating constituents and there is a risk of explosion in high temperature furnaces. Filler metals for furnace brazing do not generally contain volatile constituents so that the risk of exposure to metal fumes is small.

The most common gases used in brazing furnaces are:

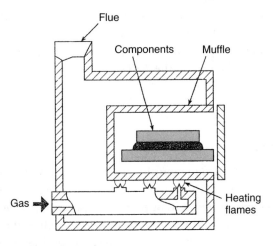

19.2   Furnace brazing.

—   hydrogen
—   cracked ammonia
—   catalytically reacted air/natural gas or air/methanol.

Mixtures of some furnace atmospheres and air are explosive and it is necessary to purge the furnace of air before it is heated. There is also a risk of explosion if the door is opened when the furnace is in operation, in the absence of the protective flame. This flame is often initiated by a pilot light, which must always be kept alight.

Continuous furnaces have a curtain of flame at the inlet and the outlet which prevents escape of gas to the workshop and this curtain should always be lit. This flame can be invisible, so it is strongly recommended that it is allowed to play on a stainless steel gauze to render it visible. It is recommended that all products of combustion should be discharged to the outside atmosphere, see Fig. 19.2.

Salt bath brazing

The components are assembled with brazing alloy and lowered into a bath of molten salt. The bath may be heated by electricity, gas or other fuel. The salt heats up the work and the brazing alloy melts; the salt also acts as a flux, removing any oxides on the surface of the work and enabling the alloy to penetrate the joint, see Fig. 19.3.

A variety of salts are used, some of which are highly toxic and appropriate exhaust ventilation should be provided. There is also a

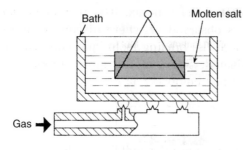

19.3   Salt bath brazing.

risk of spatter, especially when components are put into the bath: they must be dry to avoid a serious hazard. Nitrate salt baths are in common use for aluminium brazing; combinations of these salts can be explosive and care is essential, especially with dry salt. All users of nitrate salt baths should make themselves familiar with the information in the suppliers' data sheet.

Fume from the Alloy

The principal hazard in brazing, common to all processes, is exposure to toxic fumes and gases, which are generated from the constituents of the filler metals or from the coating on the metal surface. Some general notes on these fumes follow, with sections on specific fume problems which should be taken into account when work is planned.

Provided a correct technique is followed the filler metal should be melted by heat conducted from the assembly and not directly by the flame or heat source. In this way the filler metal runs without overheating and solidifies rapidly, thus giving rise to the minimum of metal fume. The main cause of excessive metal fume in torch brazing is a poor heating technique in which the operator directs the flame at the filler metal before the parent metal is at brazing temperature. Extra heat must then be applied to bring the components to brazing temperature so that the filler may run into the joint. This procedure overheats the filler metal and oxidises it, producing an unsatisfactory joint.

It is preferable to select a flux which melts some 100 °C below the melting point of the filler metal. Most manufacturers supply a range of fluxes which are suitable for use at various temperatures. Low temperature fluxes have fluorides as an important constituent and those for higher temperature have borax. It is recommended that where excessive fume is observed or where discomfort is experienced during brazing both the brazing technique and type of flux being used are investigated. Some of the more important toxic hazards are briefly described below.

Cadmium

This element is present in some filler metals, particularly those of the copper-silver-zinc type, and is frequently found in plated coatings. Cadmium oxide fume and dust are very harmful and care is necessary to ensure that they are not inhaled. If the brazing procedure is correct very little cadmium fume will be present, but even so it is important for good local exhaust ventilation to be provided except for very intermittent work in well-ventilated workshops. For production brazing, local fume extraction is necessary and for all work in confined spaces a face mask supplied with fresh air at a positive pressure should be worn.

If required, cadmium-free brazing alloys are available, but as it is necessary to use more silver to keep a reasonably low melting point, they are more expensive. They are mainly intended for use in equipment which will be used in food preparation since cadmium is also toxic if ingested.

Cadmium oxide fume is assigned a maximum exposure limit of $0.025\,mg\,m^{-3}$, averaged over an 8h period or $0.05\,mg\,m^{-3}$ for a 15 min period under the Regulations in the UK,[76] so exposure of workers must be reduced so far as is reasonably practicable and always be below the limit. The USA exposure limit is $5\,\mu g\,m^{-3}$ averaged over 8h.[170] This reinforces the case for the use of alloys with low or negligible cadmium content, unless the conditions of use can be controlled so as to ensure minimal fume generation. Brazing materials that contain cadmium must be clearly labelled in the USA, with a label similar to Fig. 19.4.[124]

Beryllium

Beryllium forms an important alloy with copper and is also a constituent of some magnesium and aluminium brazing alloys. It is highly toxic; serious illness and death have resulted from exposure to metallic beryllium and its compounds in the form of fume and dust and stringent precautions should be taken to avoid

> ## Danger – contains cadmium
>
> Protect yourself and others. Read and understand this information.
> Fumes are poisonous and can kill.
>
> - Before use, read and understand manufacturers instructions,
>   the MSDS, and your employer's safety practices.
> - Do not breathe fumes, even brief exposures to high
>   concentrations should be avoided.
> - Use enough ventilation, exhaust at work, or both, to keep
>   fumes and gases from your breathing zone and general area.
>   If this cannot be done, use air-supplied respirators.
> - Keep children away when using.
> - See ANSI Z49.1.
>
> First Aid: If chest pain, shortness of breath, cough or fever develop
> after use, obtain medical help immediately. Do not remove this
> information.

19.4   Cadmium warning notice.

contamination of the workshop atmosphere and even the atmosphere outside the factory. Precautions must be accompanied by
regular analyses of the workshop air to determine the amount of
beryllium present.

Any work on beryllium alloys must be regarded as an extremely
hazardous operation, which should not be started until suitable ventilation and other precautions are available and check analyses have
been organised. Employees should wear protective clothing to avoid
contamination of their personal clothing and close attention to personal hygiene should be encouraged by the provision of washing
facilities close to the work.

Zinc

Zinc may be present in the filler and parent metals as a coating and
the precautions outlined above for cadmium may be used as a guide.
Overexposure to zinc fume causes metal fume fever, but in the case
of zinc, recovery is usually quite rapid.

Fume from the Fluxes

Fluxes are mixtures of inorganic chemical compounds formulated to give satisfactory results for specific purposes. The data sheets supplied with them should always be consulted, to discover the precautions that should be taken in order to ensure their safe use.

Fluxes for low-temperature silver brazing are based on a combination of salts of sodium, potassium, boron and fluorine. The fumes given off when these fluxes are heated may contain small quantities of hydrofluoric acid and boron trifluoride, which are likely to be irritant to the eyes, nose, throat and respiratory passages. Consequently local fume extraction should always be provided with the possible exception of work of a very intermittent nature. Dermatitis can arise from skin contact with the fluxes and their fumes, so that precautions must be taken to avoid this hazard. A suitable barrier cream should be applied to the hands and forearms and a protective ointment of the lanoline type (Vaseline three parts and lanoline one part) may be used on the face and inserted in the nostrils several times a day. Abrasions or breaks in the skin should be immediately covered with a waterproof adhesive dressing. Washing the hands must include thorough cleaning of the nails. Fluxes for high temperature brazing have boric acid as a principal constituent and no significant ill effects arise from their use.

Where fluxes have to be mixed into a paste with water, containers must be provided which will not be confused with teacups, etc. Similar fluxes to those mentioned above are used in flux baths. Dip brazing with fluorine-bearing fluxes should be done in baths provided with efficient exhaust ventilation.

The main use for salt bath brazing is on aluminium where nitrate salts are in common use. Apart from this there are some applications in low temperature silver brazing where the salts are composed of a mixture of sodium cyanide and sodium carbonate. When higher temperature salt bath brazing is used the salts will be based upon sodium carbonate.

Local exhaust ventilation must be provided for cyanide salt baths. Cyanide salts and acids must not be stored together as their mixture will lead to the evolution of hydrogen cyanide, which is an extremely dangerous gas. Cyanide can be absorbed into the system through the skin and the salts on local contact can give rise to dermatitis so that the provision of protective clothing is advisable. Employees should adopt a high standard of personal hygiene. These salts must not be introduced into the mouth, because they are extremely toxic. Therefore meals and refreshment should not be taken at the workplace. Burns caused by splashes of molten cyanide must receive immediate first aid treatment. Special first aid treatment for cyanide poisoning should be readily available wherever cyanide salts are used. This will consist of the provision of first aiders who are specially trained to handle a cyanide emergency and oxygen-giving equipment must be readily to hand.

Surface Preparation and Cleaning Procedures

A clean oxide-free surface is imperative to ensure a sound brazed joint. All grease, oil, dirt and oxides must be carefully removed from the filler and parent metals before brazing. Unsatisfactory surface conditions can be dealt with by mechanical means such as grinding, filing, scratch brushing and various forms of machining, or by the use of a variety of chemical cleaning solutions. On occasions both mechanical and chemical cleaning will be necessary. Certain mechanical methods, such as grinding, will require operators to protect eyes with goggles. The dust produced to be removed by local exhaust ventilation.

Many different cleaning solutions are in use. Before using any preparation the manufacturer's safety data sheet should be examined, so that appropriate control measures can be taken. Cleaning solutions can be roughly classified into three main groups: solvents, acids and alkalis. Solvents should be chosen with care. Chlorinated hydrocarbons are in general being phased out and replaced by alternatives. Many of the alternatives are highly flammable and therefore the fire precautions may need to be revised.

Adequate ventilation must always be provided in areas where cleaning operations are being conducted with the use of solvents. Solvents should not be used as hand cleansers because they remove fat from the skin and render it sensitive to damage by other chemicals. Unless acid solutions are very dilute it is advisable to install local exhaust ventilation over cleaning and pickling tanks to remove acid fumes, particularly if the process is hot or electrolytic.

Operators should be supplied with protective clothing, gloves, footwear and a suitable barrier cream or lanoline ointment which can be applied to the skin of the hands, forearms, and face and if necessary inserted into the nostrils.

Hydrofluoric acid and its vapour are particularly corrosive to the skin, fingernails and respiratory passages. If it is used in concentrated form, stringent precautions are required. It is a systemic poison and can be absorbed readily through the skin when it is spilt. In the UK the HSE should be consulted. There will be a requirement to supplement the training of the first aiders to deal with any incident. Calcium gluconate gel will need to be kept readily available for the prompt treatment of burns. The use of hydrofluoric acid in dilute form must be accompanied by the preventive measures outlined above, with adequate training and warning of operators. Even small burns require medical treatment – fatal accidents have occurred when operators have been burnt and not realised the serious nature of the burn.

It is dangerous to inhale the 'fumes' from concentrated nitric acid, therefore any spillage should be hosed away to a drain with large quantities of water by persons wearing efficient breathing apparatus. Nitric acid spillages should never be absorbed with a cotton mop, cloth or sawdust.

The use of caustic alkali solutions in tanks or otherwise for cleaning metal assemblies demands the provision of similar preventive measures to those recommended for acids. Assemblies that have been brazed, particularly if a flux has been used, may need cleaning, either by grit blasting or by scrubbing by hand in a tank of hot water. The water will become contaminated with irritant flux residues, thus skin contact must be avoided.

Facilities for prompt first aid treatment of splashes of alkali or acid on the skin, or especially the eye, must be immediately available.

Physical Hazards

The physical hazards encountered in brazing are somewhat similar to those experienced in welding. In the operation of salt baths, precautions must be taken to protect the operators against spatter, which can arise from a number of causes. The most common of these are the sudden escape of entrapped air from an assembly or from hollow tools, the rapid evaporation of entrapped moisture remaining as a result of inadequate preheating and foreign bodies dropping into the bath. Particular care should be taken to ensure that workpieces are not dropped into the bath or left lying at the bottom.

When brazed aluminium assemblies are cleaned in tanks of caustic solution, hydrogen may be evolved by chemical reaction producing a spray or mist of caustic liquid. This will cause irritation, or worse, of the eyes, nose, throat and skin if it is allowed to come into contact with them. The tanks should therefore be fitted with a system of local exhaust ventilation. Employees must be provided with goggles, protective clothing and footwear, and a suitable barrier cream for the skin.

20

Soft Soldering

Soldering is a lower temperature joining process than either brazing or welding. A filler metal is used which forms a bond with the parts to be joined. In order to achieve this bond the surfaces must be clean and this is achieved by abrasion and degreasing (where necessary) followed by fluxing to remove the oxide layer on the metal to allow the liquid solder to flow and make intimate contact with the parts to be joined.

There are many commercially available alloy systems. These include:

- tin: tin–lead, tin–lead–antimony, tin–lead–silver
- lead-free solders: tin–antimony, tin–copper.

Processes

Wave soldering

Electronic components are assembled on to a printed circuit board by passing their leads through the board and flux is applied. A pump circulates solder through a bath, forming a higher 'wave', which contacts the underside of the board as it moves along the bath, soldering all the joints in one pass, see Fig. 20.1.

Vapour reflow soldering

Surface mounting electronic components with solder pad connections are attached to a printed circuit board by adhesive. The assembly is lowered into a vapour bath, which heats it and reflows the solder to make all the joints, see Fig. 20.2.

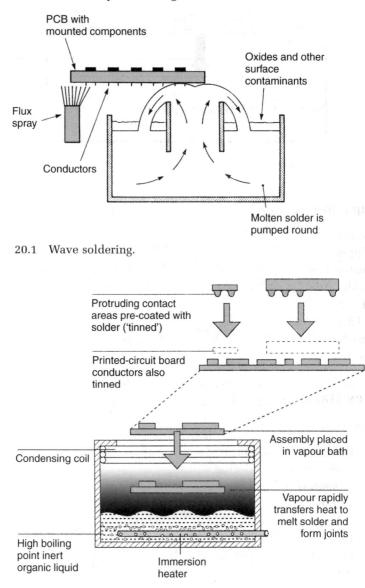

20.1   Wave soldering.

20.2   Vapour reflow soldering.

Manual soldering

This uses a soldering iron or a gas torch to heat the solder and form the joint. The choice of heating method will depend on the application, a soldering iron being adequate for electronic components, but a gas torch for general engineering.

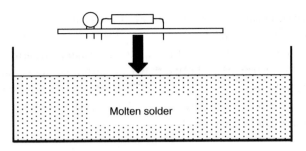

20.3   Dip soldering.

Dip soldering

Components are placed and fluxed and then the whole assembly is dipped in a pot of molten solder. The pot may be an electrically heated cast iron. The surface of the solder must be kept clear of oxide. Adaptation to wave soldering ensures that this happens, see Fig. 20.3.

Other means of melting and reflowing a preplaced solder include resistance and induction heating, oven and hotplate soldering and the use of a laser.

Key Hazards

Key hazards involve:

— the heating medium
  electrical – soldering irons
  gases – if using gas guns
  liquid baths
  laser heating.
— fumes from the filler and fluxes
— cleaning agents, for cleaning before/after processing
— fire.

Electrical hazards

These are easily avoided by using equipment that is well maintained, and operating at reduced voltages. Soldering irons should be purchased from a reputable supplier. As portable equipment, they can easily become damaged, so should be inspected frequently and tested formally at intervals. Alternating current electric soldering irons of normal voltage must be earthed, unless they are double insulated. Where it is desired to avoid earthing, for example when working on certain live circuits in electronic equipment, a low voltage (12–24 V) iron may be used with a suitable safety isolating transformer. Where radio frequency (RF) heating is used for soldering, the safety precaution previously described under brazing in Chapter 19 should be observed.

Gas heating

When a gas torch is used to heat the solder, the precautions described in Chapters 4 and 10 should be taken with respect to the gases.

Liquid baths

In vapour-phase reflow soldering, the vapour above a bath of heated liquid condenses on the work and heats it above the melting point of the preplaced solder. Special proprietary liquids are used, which are generally of low toxicity. Vapour release is nevertheless controlled to as low a level as practicable, mainly because the liquid is expensive to replace. As decomposition may occur at any hot spots on the heating element or due to any contaminant which enters the bath, moderate local exhaust ventilation is installed to ensure safe working conditions. Smoking should not be allowed in the vicinity, as vapour inhaled through a cigarette might form harmful products.

The bath of molten metal can spill or splash. Wear eye protection and shoes, and prevent any water or damp material from coming into contact with the hot solder.

Dip soldering entails the use of considerable quantities of molten solder. Articles and tools covered with flux have to be immersed slowly in the molten metal to give the flux time to boil off gradually. It is advisable to provide splash guards around the rim of the dipping bath, and protective goggles should be worn.

The flux boils off slowly in contact with the molten metal. But if a moist article is rapidly immersed in molten solder the moisture trapped on its surface evaporates, often with explosive force, leading to the metal 'spitting'. Unless the bath is covered with flux, any article dipped in the molten metal should be dried by preheating and immersed slowly and carefully. Similarly, if molten solder is poured into a mould when a solder bath is emptied, the mould must be quite dry.

Laser heating

Microelectronic assemblies, such as attaching chip capacitors to thick film circuits, can be made by melting preplaced solder with a laser, the hazards of which are outlined in Chapter 18.

Fume

Just as with welding, the exposure of personnel to substances that are hazardous to health must be controlled, to within any exposure limits (Chapter 5).[65–71]

Hazards from the filler metal

The soft solders in common use are alloys containing varying proportions of lead and tin. Lead poisoning can arise from the inhalation and ingestion of metallic lead and many of its compounds, particularly in the form of fume or dust. There is little risk to health in handling alloys in solid form, such as solder sticks or wires, provided the operators observe good personal hygiene – washing their hands before eating and drinking. The soldering process produces little airborne lead fume.

However, the inhalation or ingestion of finely divided solder, such as powder or filings, may constitute a hazard. The operation of sanding or filing down solder deposits on motor car bodies should be confined to a ventilated booth to prevent the spread of dust to other parts of the workshop and the operator should be adequately protected by a helmet supplied with air at a positive pressure.

Exposure of workers to lead is controlled by regulations.[67,68] Food, drink and tobacco must be kept away from the soldering bench and operators must wash their hands and arms before taking food. Ideally, food and drink should be consumed in a special room provided for this purpose.

Where clothing may become contaminated by lead dust it will be necessary to provide protective clothing which can be laundered under suitable conditions to prevent the spread of contamination. Lead is more toxic to young children than to adults so the operators should preferably not take contaminated clothing home to be washed.

Dross skimmed from melting or dipping pots which contain molten solder should be handled with care so that it does not become airborne and it should be collected and kept in containers with well-fitting lids. Keeping solder dross in sacks is dangerous. No dross should be allowed to collect on the floor. In many instances the dross from solder pots will be mixed with residue from soldering flux; it is then in the form of a moist sludge and although there is no risk of it becoming airborne, the substance is still toxic and should not come into contact with food or drink.

Whereas in normal circumstances soft solders are not liable to be heated to temperatures at which the formation of fume takes place, it is possible for solder to be spilled onto glowing coke or charcoal if the solder pot is heated on a brazier: this practice should not be used. Cadmium-containing solders (such as cadmium–silver solder and cadmium–zinc solder) should be handled carefully and employees should avoid exposing themselves to fume.

Hazards from the flux

Fluxes used in soft soldering fall into two main groups: inorganic and organic fluxes. Inorganic fluxes may be acidic (containing phosphoric acid, for example), alkaline (containing amines and/or ammonia) or salts (containing ammonium chloride or zinc chloride). They are highly active and are effective on surfaces that are difficult to solder. However, they are highly corrosive.

Zinc chloride is a corrosive substance and is toxic if taken internally; it should not be allowed to contaminate food or drink. It is a skin irritant and can cause dermatitis. The introduction of zinc chloride into the eye will give rise to a severe inflammation. Zinc chloride fumes may cause ulceration of the nasal passages. In most circumstances it is sufficient to insist upon good housekeeping practice and to instruct operators to avoid contaminating their skin or garments with the flux. Some operations which demand the use of large quantities of flux, such as the dip soldering of radiators, give rise to excessive evolution of flux droplets into the workshop atmosphere.

Whenever an active flux is used, a barrier cream, supplied on medical advice, should be applied to the hands and forearms. The use of protective gloves should be discouraged because, once they become contaminated with zinc chloride on the inside, they are difficult to clean and promote, rather than prevent, dermatitis. Dermatitis is often caused by operators using dangerous materials such as solvents to clean their hands, not through the action of the flux. It is impossible to remove active flux from the hands with ordinary soap because it forms an insoluble compound with the flux. A degreasing agent, such as petrol must not be used to wash the hands because it dries and cracks the skin and then contact with an active soldering flux is painful and can lead to ulceration of the skin and to severe dermatitis. Safe hand cleansers which remove flux efficiently and do not lead to skin damage must be used; medical advice may be needed to select a suitable formulation. Fluxes intended for soldering stainless steel and similar metals may contain hydrofluoric acid and should be treated with special care: observe the supplier's recommendations.

Active fluxes are eye irritants. Where working conditions make it likely that flux will splash, safety goggles or screens should be used.

Safety fluxes

So-called 'safety fluxes' are based on 'rosin' or 'colophony', pine sap tapped from various species of pine tree. In its solid or semi-liquid form it can act as a skin irritant: this is unlikely to occur in organised production, as parts should be untouched by hand, to avoid contamination and failure of the solder to wet the surface. If flux does get on to the skin, a safe hand cleaner should be used, as soap is not effective.

It is now apparent that colophony is a respiratory sensitiser. A proportion of the population exposed are liable to develop an allergic asthma.[63,64] Precautions should be taken to control the exposure of the workforce to fume. It is advised that medical surveillance is instituted to detect early signs of sensitisation, so that the exposure controls may be reviewed. Workers should be informed of the measures that they should take to avoid exposure.

When the flux comes into contact with the molten solder, it partly decomposes to give off a range of fumes. These may cause eye irritation in the short term. The exact constituent or constituents of the fume have not been positively identified. As rosin is a natural

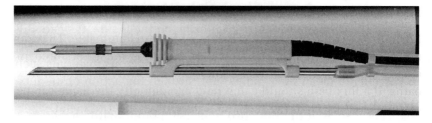

20.4   Soldering iron with integral fume extraction (photograph courtesy of Antex Ltd).

product, the amount of fume and its composition may vary, so it is not possible to fix an objective measurement to indicate satisfactory fume control, unlike other airborne pollutants affecting the respiratory system.

The majority of electronic work today is machine soldering of printed circuit boards (PCBs). The machines are normally installed with overhead extraction hoods to remove all fumes before they can reach any worker's breathing zone. For production line manual assembly work, some form of effective fume control is essential. Soldering irons are commercially available (Fig. 20.4) with a fume extractor tube extending to just above the bit. The tubes from up to ten irons are connected through pipework to a filter collecting solid residues and a pump. With a high workload, supplementary overhead extraction may be required.

Cleaning agents

Surfaces must sometimes be cleaned prior to soldering. To avoid corrosion of the soldered job it is usually necessary to wash off the flux. Several agents may be chosen: acids can cause serious burns and the vapours are highly irritating to the nose and respiratory system. Alkali baths may contain caustic soda, trisodium phosphate or other soaps, wetting agents or emulsifiers. If caustic soda is left on the skin it can produce deep painful burns. These materials present a serious hazard if splashed in the eyes. Organic solvents can remove natural oils, and breathing the vapours can have various harmful effects.

Employers should read the data sheets from the manufacturers of the substance that they choose to use and put in place measures to prevent exposure. Wash water should be disposed of safely.

Fire hazards

Rosin fluxes contain volatile solvents which are flammable. They do not constitute a high fire risk, but all supplies must be stored in closed containers. Naked lights must be kept away from open flux baths. If a flux bath should catch light accidentally the flame can be extinguished by covering the container with a sheet of metal.

21

Thermal Spraying

Thermal spraying is widely used in industry, both to reclaim worn parts and to deposit a surface coating to impart properties such as corrosion or wear resistance. Spray materials include ceramics, metals, polymers (see Chapter 22) and mixtures.

Only general principles are outlined here, as it is not possible in this short account to deal with all the many processes and materials used. There is an Industry Code of Practice that contains much more information.[171] If materials not mentioned in this chapter are to be used, it will be particularly important to obtain information from their manufacturer or supplier concerning any hazards their use may present, whether in terms of toxicity or flammability.

In thermal spraying, coating material in the form of wire or powder is fed into the nozzle of a gun where it is fused by the heat source and projected as a spray onto the surface to be coated. The heat source can be a fuel and oxygen, an arc, or a plasma.

Flame Spraying

Flame spraying guns are light and easy to manipulate. They burn acetylene, propane or hydrogen with oxygen and the spray consumable is in the form of a wire, cord or powder. An additional gas jet is frequently incorporated to accelerate the particles towards the workpiece.

Arc Spray

In this process, the heat source is an arc, struck between two wires, which are both consumable. A jet of gas is used to project the molten metal droplets towards the surface to be coated. This process is

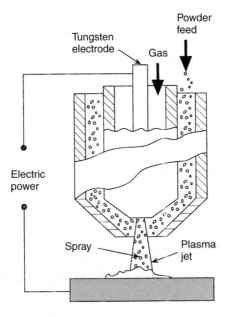

21.1   Plasma spraying.

capable of use on site, for example for spraying bridges with zinc or aluminium.

Plasma Spray

This process uses a plasma to act as the heat source. The consumable is fed in as a powder and ejected at great speed towards the workpiece, Fig. 21.1.

High Velocity Oxy-fuel Spraying (HVOF)

This process uses a fuel–oxygen mixture, but is distinguished from the flame spraying system by the high velocity of the flame – up to $2000\,\mathrm{m\,s^{-1}}$. Fuels include propane, acetylene, hydrogen, propylene and liquid fuels such as kerosene. The consumable is in the form of a fine powder. The surface of articles to be sprayed will often be prepared by degreasing and or shot blasting. General advice on the safe choice and use of solvents or pickling solutions for degreasing is given in Chapter 23.

Table 21.1. Key hazards in the main spraying processes

Hazard	Flame spray	Arc spray	Plasma spray in air	Plasma spray in vacuum	HVOF
Fire and explosion	*		*** (due to use of hydrogen)	*** (due to use of hydrogen)	***
Noise	**	**	***	**	***
Fume and dust	* to ***	* to ***	* to ***	* to ***	* to ***
Electric shock	*	*	*	*	*

* Lower level of hazard.
** Medium level of hazard.
*** Very high hazard.

Key Hazards of the Processes

Table 21.1 gives an indication of the level of hazard in the thermal spraying processes. However, potential users must assess their own situation carefully.

Fire and explosion hazards

The greatest risk of fire in flame spraying and HVOF comes from the use of highly flammable gases such as acetylene, propane, hydrogen, etc. The storage of these gases should be in accordance with the advice given in Chapter 4. In thermal spray, gas flows are large and thus bulk supplies are more likely to be required. Advice should be sought from the gas supplier in setting up a bulk store and delivery system. For an HVOF system, a hazardous area classification will be required which will define zones in which electrical equipment must be to a standard to be safe in a potentially flammable atmosphere.[49,172] The equipment should use approved hose and connectors (see Chapter 10) and be provided with non-return valves and flashback arrestors to protect the system against faults.

Many of the processes use consumables in the form of powders. Many metal dusts, such as aluminium and titanium are a significant fire and explosion hazard. It is essential therefore to remove metal dust from spray booths or other confined spaces by adequate exhaust ventilation. All ducting for ventilation and dust collecting plant should be provided with blow-out panels and all equipment used for ventilation, including motors fans and pipes should be electrically earthed.

Explosion reliefs should not discharge into workrooms, therefore dust conveyors or ducting should be as short and strong as possible, or be sited outside work rooms. If it is found necessary to repair, weld or flame-cut such ducting, it should be thoroughly washed down and all metal dust removed, because the application of a torch to a ferrous base covered with a coating of certain dusts causes an exothermic reaction to take place, which damages the base and prevents proper welding from being carried out.

Material that has failed to stick to the workpiece may be collected by a wet wash system or a dry collector. While wet collectors appear to be attractive because of their capacity to suppress dusts, they carry a risk of legionellosis if there is the potential to produce droplets. Dry collectors should be sited outside the building where possible and be provided with explosion relief. A fire trap is required on the incoming side where sparks and hot metal may be a problem. Cyclone-type collectors should also be located outside the building, preferably on the roof. It is well known that certain metal dusts, when partially wetted, are capable of spontaneous combustion, so the cyclones should be protected against the entry of moisture. For good ventilation practice it is always desirable to consult properly qualified ventilation engineers. Figure 21.2 shows the external structures that are part of a fume handling system.

When cleaning out the booths, ductwork and cyclones (it is advisable to do this regularly to prevent excessive dust accumulation) the ventilation fans should be left running, all sources of ignition in the area should be eliminated and non-sparking tools should be used for cleaning and repair work. The structural steel work in the spray shop should be cleaned down regularly, particularly on the top sides of girders where most of the metallic dust accumulates, otherwise a damaged electrical cable or some other electrical defect may ignite the deposit.

A secondary explosion may follow if, for instance, a minor disturbance in the building causes vibration, shaking the dust lodging on the beams, in turn giving rise to a dust cloud of explosive concentrations within the work room for a small instant in time. This cloud might be ignited and an explosion could then be generated on a massive scale. Efforts to prevent this sort of occurrence must be directed towards the prevention of dust accumulations in the first place by:

- The proper design of buildings housing the hazard, for example, walls with smooth surfaces, and where practicable the minimum

21.2   Fume extraction system (photograph courtesy of
Metallisation Ltd).

number of ledges and obstructions on which dust accumulations
are possible.
– The modification of existing buildings, including fittings to
prevent such dust accumulations: for example, boxing in girders
and the removal of unnecessary fittings.

Aluminium dust must not be collected in the same receptacle as
iron dust from blasting or spraying. If any metal dust catches fire, it
can be difficult to extinguish. Suitable powder extinguishers of a
type rated for use on metal fires should be available nearby.

A thermal spray must not be directed towards a container which
holds or has held combustible materials and spraying should not be
carried out in close proximity to these materials.

Noise

Spraying guns, with the exception of flame spray guns, are noisy.
For instance, noise levels over 130 dB(A) have been recorded close

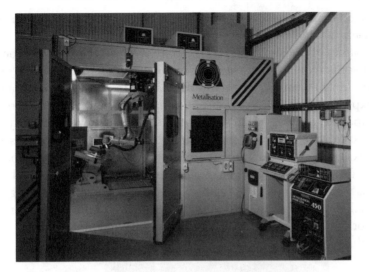

21.3   Sound attenuated spraying booth (photograph courtesy of
Metallisation Ltd).

to an HVOF system. It is desirable to house the complete equipment
inside a sound attenuated booth, see Fig. 21.3. This booth can then
be engineered to provide the necessary ventilation for the removal
of fume. The control panels are on the outside and the operator may
view the process through windows in the booth. If manual operation cannot be avoided, the operator and others exposed to the noise
must be provided with adequate hearing protection (See Chapter 8).

Toxic hazards

Coating materials

The greatest single application of the thermal spraying process is in
coating ferrous metal with zinc or aluminium to prevent rusting.
There are many other important engineering uses of thermal spraying, ranging from the reclamation of worn parts by building up with
a suitable metal, to the application of numerous types of wear-
resistant coatings. Spraying complete electrical circuits by using
special stencils is an example of the more unusual uses of the
process. Any metal, ceramic, plastic, resin, wax, etc, which melts
without decomposition can be applied by thermal spraying techniques. Some of the other metals or alloys that are used are tin,

cobalt and tungsten carbide mixtures, copper, brass, bronze and steel. A wide variety of ceramics may also be sprayed.

Many of the substances commonly sprayed are known to be toxic to some extent when they are inhaled in the form of fume or very fine dust – some of these are dealt with in the tables in Chapter 5. Exposure to these dusts must be controlled to prevent workers being exposed to excessive concentrations. Even if not specifically toxic, dust in general must also be controlled.[66,76] The user must assess the risks by undertaking research into the properties of the materials to be used and implement suitable control measures. Control in the case of mechanised spraying is best achieved by working in an enclosure or booth with suitable exhaust ventilation. For manual work it may be necessary to equip the operator with an air fed helmet.

Plastics, resins and waxes

The powder process may be adopted to apply coatings of certain plastics resins and waxes by a simple modification of the spraying procedure. Some of these materials give rise to unpleasant fumes which are irritants to the eyes and respiratory passages; consequently the operator must be given adequate personal protection. The fumes must be removed by an efficient local exhaust ventilation system and discharged into the outside atmosphere at a high level.

Surface preparation by blasting

Blasting is carried out with abrasive materials such as chilled iron grit, steel or aluminium oxide grit. Sand or other substances containing free silica must not be used, as anyone exposed to dust from it could develop silicosis.[76]

In a factory, blasting operations should be carried out in a suitable enclosure or room to protect other personnel from injury and nearby machinery from damage. The 'blast room' should be provided with an efficient system of exhaust ventilation, preferably of the downdraught type. During the blasting operation abrasive material rebounds from the surface of the article with a high velocity. Consequently the operator must be given special protective clothing such as gloves, apron and leggings. A helmet supplied with fresh air at a positive pressure is also necessary to protect the blaster from both flying particles and harmful dust.[82]

Because of the friction between the finely divided particles of grit and the blasting hose and nozzle, discharges of static electricity occasionally take place. It is advisable to earth the blasting hose and nozzle.

Physical and electrical hazards

A thermal spraying gun must never be directed towards any person. Protection of the eyes and skin of the operator should follow the requirements of the equivalent gas, arc or plasma welding process described in earlier chapters. Where the arc or plasma stream is shielded from the direct view of the operator, a lighter filter glass will allow a clear view of the work. Arc and plasma processes involve similar dangers from electric shock as the equivalent welding processes. The safe practices prescribed in the appropriate chapters should be followed.

Workpieces are frequently mounted on rotating tables or X–Y tables. The thermal spray gun is also frequently mounted on a traverse. The thermal spray booth, which serves to reduce the noise and provide fume extraction, also serves to protect the operator from mechanical hazards. Any 'teaching' of the system should be done at reduced speed.

22

Welding and Flame Spraying Plastics

Plastics are well established as engineering materials and their applications continue to increase as improved materials and methods of manufacture become available. Among these methods are a number of welding processes, some apparently similar to the processes for welding metals. Plastics materials are divided into two main classes: thermoplastics, which soften and flow on heating, and thermosets which do not. Only thermoplastics can be welded.

Plastics soften gradually as their temperature is raised and do not form a molten pool as in the welding of metals. This means that there is no problem of spatter from a weld pool. However, if overheated, plastics materials will decompose irreversibly. Since successful welding depends on the avoidance of such overheating, there will be little fume generated in normal operation.

Welding Processes

Hot gas welding

Hot gas welding is carried out with a gun which is used to direct a stream of gas at a temperature which is usually in the range of 200–400 °C. The most common type of gun is electrically powered with a fan propelling air through a heating element; electronic control ensures a consistent temperature, see Fig. 22.1. For some applications nitrogen gas from a cylinder is preferred. Most thermoplastics can be welded by this process.

The minor hazards presented by the process can be avoided by:

– safe handling of gas cylinders as explained in Chapter 4
– correct installation and maintenance of electrical equipment

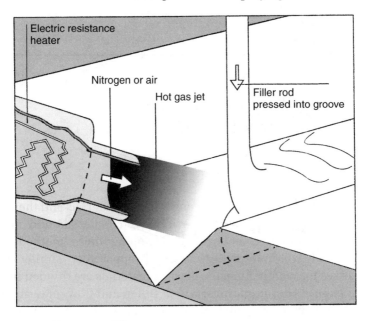

Electric resistance heater

Nitrogen or air

Hot gas jet

Filler rod pressed into groove

22.1   Hot gas welding.

– good fume extraction if working in a confined space, especially when another gas is used instead of air
– keeping hands clear of the hot gas stream.

In most cases the use of gloves is not necessary.

Heat sealing

Where heat sealing is performed with a hand or foot operated clamping action, hazards will be minimal. If power clamping is used, protection against the operator's hands being trapped should be provided by a guard or two-hand button system, see Fig. 22.2.

Hot plate welding/heated tool welding

Hot plate or heated tool welding finds many applications, both in factory production of such items as the cases of automobile batteries and on site, for example, in joining gas service pipelines in highway excavations. A hotplate is electrically heated to a temperature which may be as high as 350 °C. The parts to be joined are clamped on either side of the hotplate. When the parts are hot

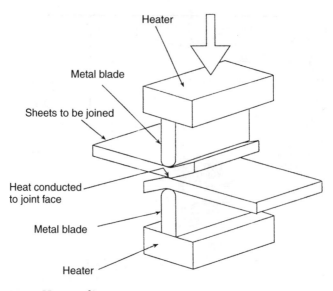

22.2   Heat sealing.

enough, the hotplate is withdrawn and the parts clamped back together to make the joint, see Fig. 22.3.

Care is clearly needed to avoid burns to the hands, but guides, etc, will often be built into the equipment to allow a swift changeover from heating to welding actions. Again electrical equipment must be properly installed and maintained.

Dielectric welding; also high frequency welding (HF welding); radiofrequency welding (RF welding)

A typical machine for HF welding of plastics consists of a press which is provided with a fixed and a moveable system of electrodes between which the article to be welded is clamped at the time of operation. Power is supplied to the electrodes from an HF generator and frequencies of around 20–150 MHz are most commonly used to weld thermoplastic materials such as plasticised PVC and polyurethanes, see Fig. 22.4.

The principal hazard of HF welding is one of contact by the operator with the live electrode, which is likely to result in an HF burn. Such a burn can be very painful as it tends to penetrate deeply into the tissues, affecting tendons and nerves; subsequently, healing may take a long time. The welding machine should be supplied or fitted

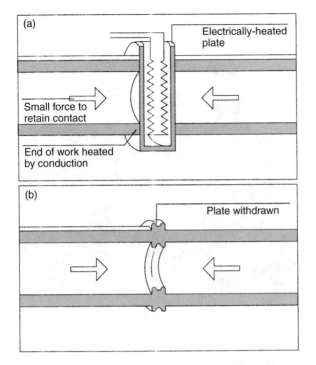

22.3   Hot plate welding (a) heating, (b) forging.

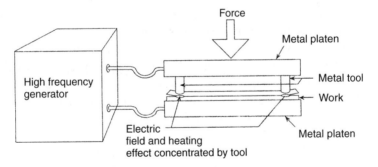

22.4   Dielectric welding.

with adequate guards to prevent access to the live parts of the system.

Because the machine often has a press action, guards also serve to prevent the operators hands being trapped (Fig. 22.5). Care is needed to maintain this protection where the user builds or modifies tooling, or when covers are removed for maintenance or adjustment.

22.5   Hydraulic 10 tonne press and 30 kW high frequency generator with guards.

Friction welding; spin welding; vibration welding

The hazards here are mainly those common to all rotating machinery. Guards will be needed if there is a possibility of access to dangerous moving parts. Vibration welding may generate sufficient sound energy to require the working parts to be confined to a suitable enclosure to prevent excess exposure of workers to noise.[106]

Ultrasonic welding

The normal operating frequency of ultrasonic generators for welding of plastics is between 20 and 40 kHz. This is just above the normal audible range so that most work on a small scale will not require a sound-absorbent enclosure. Work can sometimes vibrate at subharmonic frequencies, within the audible range; with larger equipment utilising multiple heads and power clamping, protection against the operators hands being trapped and effective attenuation of the high frequency sound can be provided by an enclosure of relatively thin metal sheet with a layer of vibration damping material, see Fig. 22.6.

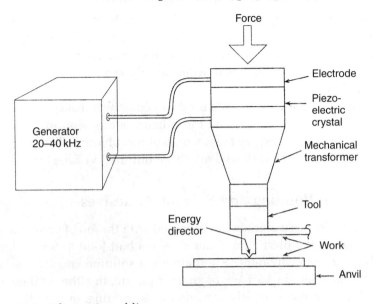

22.6   Ultrasonic welding.

Flame Spraying Plastics

In the main, the equipment for flame spraying plastics is identical to that for flame spraying metals. The surface to be sprayed is normally prepared in a similar manner, by shot blasting. Most problems in health and safety while flame spraying plastics have already been covered by the precautions suggested for the handling of fuel gases and oxygen in earlier chapters. However, certain additional possibilities should be borne in mind:

- Molten thermoplastics may adhere to the skin and cause deep burns.
- On oxidation, chlorinated polyether, polyethylene polysulphide, epoxy resins and cellulose acetate butyrate give off pungent vapours with a disagreeable smell.
- Plastics containing chlorine may give off harmful acidic vapours when heated in the presence of air.

Personnel will not be affected by obnoxious and harmful vapours if flame spraying is carried out in an adequately ventilated enclosure.

None of the thermoplastics that can be readily flame sprayed presents a serious fire risk if the equipment manufacturers' instructions for the use of the equipment are followed correctly. Finely divided polythene and epoxy resins can ignite in the flame of a spray gun if the rate at which powder is fed into the gun is too high. Under these conditions a flame may be thrown for several metres. Prompt action, in first shutting off the powder supply then the gas supply, will eliminate the danger.

Some of the plastic dusts used can constitute a class 1 dust explosion hazard. Overspray and the exhaust equipment provided for its disposal may also give rise to an explosion hazard. The precautions which should be followed are those outlined in Chapter 21.

Solvent Bonding and Solvent Adhesives

Many plastics materials can be joined with the aid of a solvent which dissolves the plastic on either side of a butt joint to form a bond as the solvent evaporates. Alternatively a solution can form an adhesive between the surfaces of another plastic. In either of these cases, which are not normally considered as welding, evaporation of the solvent produces a substantial vapour concentration with possible hazards of toxicity or flammability. The potential exposure of employees to such vapours must be assessed and arrangements made to control exposure where necessary.

Fume Hazards

The majority of thermoplastic materials which are welded for engineering applications are highly polymerised and are comparatively inert; therefore they are unlikely to give rise to any serious health hazards unless they are partially decomposed. Before welding any unfamiliar plastic material it is advisable to consider the possibility of a toxic hazard from gases and fumes, so that adequate precautions can be instituted if these are necessary.[65,66] Information may be sought from the manufacturers or suppliers of the plastic. For instance, overheating of PVC results in the production of small quantities of hydrogen chloride gas, although the concentrations are unlikely to reach hazardous proportions in an adequately ventilated workshop.

At temperatures in excess of 250 °C, polytetrafluoroethylene (PTFE) decomposes with the evolution of toxic fluorine compounds. Inhalation of PTFE can produce symptoms similar to those experienced during an attack of metal fume fever. This material must not be welded unless exhaust ventilation is applied as near to the weld as possible. In addition smoking in the workshop must be prohibited and smoking materials should not be taken into the workshop to avoid their contamination with PTFE dust and fume. Employees should be provided with overalls and good washing facilities adjacent to their workplace.

23

Inspection and Testing

A range of techniques is used to examine welds after completion, using both destructive and non-destructive methods.

Radiography

The examination of welds by ionising radiations from X-ray tubes or from sealed radioactive (gamma) sources is common practice. Ionising radiation presents a hazard which differs from many others, such as fire or injury, in a number of respects:

- It cannot be detected by human senses.
- The effects are delayed, often by many years.

The health effects of radiation were described in Chapter 7.

Sources of gamma radiation are purchased rated by their activity, in becquerel (Bq), which is a measure of the number of disintegrations per second, or curies (Ci). The curie is equivalent to 37 GBq.

Avoidance of hazards

The International Committee for Radiation Protection (ICRP) recommends three steps to avoid any significant hazard to any individual or to the population at large. First, persons should only be exposed to man-made radiation where it can be justified by a significant benefit, which applies in the case of industrial radiography. Second, exposure should be reduced to a figure that is as low as reasonably practicable, signified by the acronym ALARP, or as low as is reasonably achievable, ALARA. Third, maximum permissible doses which should not normally be exceeded have been fixed; for example, in the UK, whole body doses in any period of one year should not exceed 20 mSv for employees over 18 years old, 6 mSv for employees under 18 years and 1 mSv for members of the public. The general dose limit in the USA is 1.25 rem (roentgen equivalent man) per quarter. Further information on the dose limits is in the literature.[103,104]

The employer will need to plan the work using ionising radiation to reduce the probability of anyone receiving a dose to a very low level. This will include:

- shielded enclosures
- interlocked doors
- search routine to ensure no-one is left inside
- panic buttons inside enclosures
- audible/visible warnings of impending exposures
- personal dosimetry
- training
- written procedures
- emergency plans.

Employees who use gamma ray sources will normally be classified workers, subject to medical surveillance and have their dosimetry records kept by an approved laboratory. The employer must also keep records.

Sources for radiography

The source of ionising radiation for radiography may be either an X-ray tube or a radioactive isotope. An X-ray tube will only emit radiation when power is applied. Hence it is a reasonably straightforward matter to interlock the power supply to the X-ray tube to the doors of the radiation bay. Therefore, provided the bay is searched before closure, no-one should ever be inside when the X-rays are being emitted.

Gamma ray sources or isotopes

An isotope emits gamma rays continuously, so it can only be handled in a heavy shielded container, which is equipped with a mechanism to move the small capsule containing the isotope into a working position outside the shield, or to move part of the shielding away, for the required duration of exposure. The isotope itself is sealed in a capsule to prevent any escape of radioactive dust; the sealing should be checked regularly by a suitable technique. The capsule should never be handled directly; tools or remote handling devices should be used so that the operator stays as far away as possible from the source, thus minimising exposure. The containers are constructed to an international specification which ensures three things: that they will resist damage, for example from accidents in transport, that there is a clear indication whether the source is exposed and that they can be locked in the safe position.

Sources should preferably be stored in their working containers, otherwise a special secure store must be constructed, with adequate shielding, usually being sunk in a pit in the ground. If a source is transported by road or rail, the container must be locked in the safe position, monitored for radiation, labelled with the prescribed label and adequately secured against movement. Radioactive substances are hazardous. Persons transporting them must comply with any regulations regarding their transport.

In the UK, the HSE must be notified 28 days before any radiography is to be carried out, or a source acquired. The employer will require the assistance of a radiation protection adviser and radiation protection supervisors. The employer will also require a licence to hold the radioactive substance from the Environment Agency. Records must be kept of all sources and their regular inspection and testing. Their location should be advised to emergency services.

If a source fails to return to a safe position after use, radiographers should be aware of action to take. This will initially be to remove and to keep out all except classified persons from allareas subject to excessive radiation; then, unless they are competent to conduct the recovery operation themselves, they should call for skilled assistance. Employers must ensure that assessment is made of the consequences of any reasonably foreseeable accident, occurrence or incident, and that they should take reasonable steps to prevent and limit such occurrences and inform and train employees.

Monitoring the dose

Dosemeters measure the total exposure over a period. There are various designs, some of which contain photographic film and some of which contain a thermoluminescent material. The dosemeter is in the form of a badge that is pinned to clothing, and incorporates filters to help differentiate between possible sources of excess radiation. It is returned on a regular basis, for example every month, to an independent approved laboratory for measurement and recording.

Another monitor which measures total dose over a period, but giving an instantaneous readout, is the quartz fibre electroscope in which a charged capacitor is discharged through an ionisation chamber with the remaining voltage measured by the electrostatic deflection of a quartz fibre. Several electronic devices are also available that measure cumulative dose in microsieverts, µSv. These are extremely useful when carrying out hazardous operations such as changing the source, because the dose received can be read out immediately.

Special enclosures

Radiography is best done inside a clearly identified 'controlled area' to which access is controlled by walls and doors with interlocks and warnings, see Fig. 23.1. On commissioning an enclosure, tests are made to ensure that radiation outside the enclosure does not exceed the dose limits. Where persons such as crane drivers are working above, it may be necessary to shield the roof. Warning signs, incor-

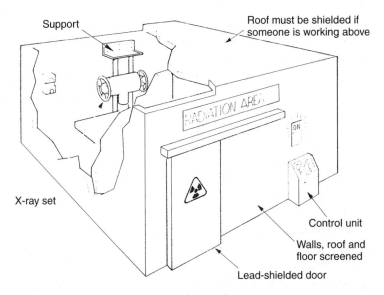

23.1   Special enclosure for radiography.

23.2   Ionising radiation warning sign (black on a yellow background).

porating the international radiation warning symbol (Fig 23.2) and lamps should be placed by the entrance. There should be an alarm and emergency stop button for the use of anyone inside when exposure is about to start; any workers who are likely to work in the enclosure such as fork lift drivers must be aware of what to do. A monitor must confirm that gamma ray sources have been safely returned to their containers at the end of an exposure.

Site radiography

Site radiography is a special case, because the protection from radiation is entirely derived from procedural control. It should be exceptional – if the object to be radiographed can be moved to a purpose-built enclosure, then it should be taken there. The authorities will require prior notice of site radiography.

Where it is not possible to carry out radiography within a permanent enclosure as described above, radiation is restricted as far as possible by shielding (Fig. 23.3). A temporary controlled area is marked out by barriers or tapes carrying the appropriate warning sign so that radiation at or outside the barrier, as measured by a survey monitor, does not exceed the required dose limit. Until work has finished only classified workers may enter the area. There should be at least two classified persons working together; while exposure is taking place, they must oversee the area. If the controlled area is large, it may be necessary to take radiographs outside normal working hours.

The removal of hand tools and any personal property of other workers from inside the controlled area eliminates a temptation to illicit entry. Welders should be made aware of the dangers of ionising radiation, how the hazards are minimised by the procedures

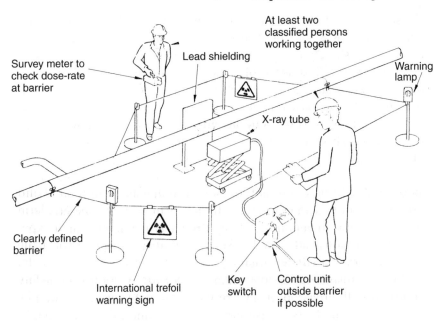

At least two
classified persons
working together

Survey meter to
check dose-rate
at barrier

Lead shielding

Warning
lamp

X-ray tube

Clearly defined
barrier

International trefoil
warning sign

Key
switch

Control unit
outside barrier
if possible

23.3   Safety precautions for site radiography.

adopted, to recognise radiographic equipment and warning signs, and to respect barriers and demarcation tapes. Finally, care must be taken, when moving source containers or X-ray tubes, to ensure that their often considerable weight is safely supported to avoid personal injury or damage to equipment.

Other Crack Detection Methods

Several other crack detection methods are in common use, including magnetic particle inspection, dye penetrant inspection and ultrasound. For magnetic particle inspection and dye penetrant inspection many chemical preparations are used. These include solvents, dyes and 'developers'. Many are conveniently packaged in aerosols. Others are made up into baths, into which the parts to be tested are dipped.

Solvents, in general, have a defatting action and are therefore harmful to the skin. They may produce dermatitis in the long term unless skin contact is minimised. They are generally relatively volatile, giving off vapours that can be inhaled. The health effects depend on the exposure and the relative toxicity of the solvent.

These may include irritation of the eyes and lungs, headache, nausea, dizziness and light-headedness. Unconsciousness or even death can result from exposure to high levels of solvent vapours.

At one time, chlorinated solvents were the preferred option, since they are generally non-flammable. The strategy was to use the solvent with the least toxicity. However, owing to concerns about the atmosphere, many common chlorinated solvents are now no longer used. This has caused employers to choose alternatives, some of which are more toxic and some of which are highly flammable. Thus, before using a solvent, the manufacturer's safety data sheet (MSDS) should be examined, to gain an understanding of all of its properties, so that measures can be taken to store it correctly and avoid exposure to vapours and any fire hazards that may arise. Solvents should always be used in well-ventilated areas, to avoid high concentrations of vapour arising.

Many aerosols used to contain chlorinated hydrocarbon or chlorinated fluorocarbon (CFC) propellants, but these are now not used. The propellants in use now include some highly flammable substances such as butane. The MSDS should be read carefully. Some chlorinated solvents are still used. These tend to decompose in an arc to form extremely toxic products such as phosgene and hydrogen chloride. Therefore it is essential to ensure that solvent vapours cannot leak into areas where welding is being undertaken.

Bulk supplies of solvents or aerosols should be stored separately from the work room, in containers appropriate for their properties, e.g. highly flammable substances, and preparations should be stored in metal cabinets, clearly labelled to indicate the nature of the contents. Only the quantity required for immediate work should be kept in the work room.

Where vapour degreasing baths are used, these should be kept in good working order – the design should contain the vapour so that employees are not exposed.

Wiping objects with solvent impregnated tissues should be avoided where possible, since the discarded wipes can be a fire hazard and this practice tends to lead to high exposures to the skin and respiratory system.

Wherever there is a possibility of getting fluid droplets in the eye, eye protection should be worn. Personnel should not eat or smoke in areas where there are solvents. They should wash thoroughly after working and before eating or smoking.

Etches

It is common practice to take sections of a weld, grind the surface and examine the microstructure of the metal when it is revealed by a suitable etch. The etches in common use are frequently acid mixtures, each chosen for a particular purpose, e.g. to show up different features of the microstructure.

When mixing etches, the instructions should be followed carefully, because in many cases the order of mixing chemicals is important. Instructions should be followed regarding storage, since some etches can deteriorate in store and explode. Employees should avoid getting etches on their hands, by using gloves, tongs or other means. Eye protection should be worn. There should be eye washing facilities near to the etching station, in case of accident.

24

Welding in More Hazardous Environments

There are certain environments in which welding is significantly more hazardous than in workshop conditions. This chapter describes permit to work systems and gives advice on work in the following situations:

- Vessels that are contaminated by flammable materials
- Confined spaces
- Situations of increased risk of electric shock
- Environments containing substances hazardous to health.

Permit to Work Systems

When intending to embark on work in hazardous environments or work that is intrinsically more dangerous than in normal circumstances, it is often considered to be good practice to instigate a permit-to-work system. This is particularly appropriate where some or all of the following apply:

- There must be formal preparation for the work, such as locking-off valves and power supplies, testing the atmosphere, clearing the work area of combustible materials, etc.
- The work must be done in a certain way, such as while wearing self-contained breathing apparatus.
- There must be a formal end to the work, such as to permit the unlocking of valves, or the examination for fire.

Permits to work are valid only for a short period and are issued in person to people who are to undertake the work. They should specify what has been done in preparation to make the area safe, how the work should be done and how the final handover will be done. They should only be valid for a restricted time – normally a single shift at most. The welder's duty is then limited to understanding and observing strictly the conditions laid down, obeying the agreed handover procedure on change of shift or completion of the work, and reporting promptly any unexpected incidents which might indicate an unsafe situation, such as pipes found to be under pressure when disconnected.

Permits to work are widely used in oil refineries, chemical plants and power stations. They are also used in other workplaces, especially where 'hot' work is to be carried out.

Vessels that are Contaminated by Flammable Materials

It is dangerous to weld, cut, grind or saw into a vessel which has at any time contained a flammable liquid unless proper precautions are taken. Work should not be attempted in the absence of proper facilities to ensure safety. An explosive ignition of vapour may be caused by the arc or flame used in welding, cutting and brazing, or even by a hot soldering iron. The danger is present not only in vessels that have held volatile liquids such as petrol, but also in those which have contained liquids such as tractor vaporising oil, diesel fuel, paraffin, linseed oil, concentrated aqueous ammonia, etc. As little as a spoonful of a flammable liquid left in a drum can cause a fatal accident. Advice is given in several publications which should be consulted before proceeding.[173-175]

If a vessel has to be worked upon and its previous contents are unknown, it should be treated as if it has contained a flammable substance, however long it may have remained empty. The first consideration is whether the work could be done by a cold process, such as cutting with shears, replacing the item rather than repairing it or using a cold repair process.

Methods of preventing accidents are of two main types:

- Removing the flammable material,
- Making the material non-explosive and non-flammable.

Precautions relating to small vessels, which cannot be entered, are discussed in this section. Large vessels that the worker will enter are discussed in the following section as 'confined spaces'. It is the duty of management to provide a safe place of work, with safe access and a safe system of work, for example by appointing and training a responsible person to supervise the preparation of vessels for welding and their testing, or by organising a permit-to-work system. It is the duty of welders and other workers to carry out the orders given by management to ensure their safety, for example, to follow a permit-to-work system. They should question any unclear instruction bearing on safety and report any incidents which may indicate an unexpected hazard.

Isolating and emptying the vessel

The vessel should be isolated from other equipment. It is preferable that this isolation is by removing the pipes connecting it to other equipment physically, or by the insertion of blanking plates – relying on isolation valves can be risky, because even a small leak can lead to a serious hazard. The vessel should be emptied and residues removed so far as possible. However, this will not be sufficient to make the vessel safe. Cleaning, gas freeing or inerting will be required.

Gas-freeing

Gas-freeing or purging is the removal of flammable vapour from a container by passing forced air ventilation through it for sufficient time to remove the flammable vapours. Natural ventilation cannot normally be relied on to remove vapours. Unprotected electrical equipment must not be used until the flammable vapour has been removed.[49] Ventilation is continued until the vapour concentration at all points inside the vessel is lower than 5% of the lower flammable limit of the vapour being removed.

Gas-freeing on its own is not sufficient in most cases. If there are residues in the tank, they can still present a danger, since they may become vaporised during hot work and create a flammable atmosphere.

Cleaning

Residues may be removed by cleaning, using several methods. Steam may be passed through the vessel, taking care not to allow the pressure inside to rise. Condensate should be allowed to drain away from the lowest point, so that residues can drain away, see Fig. 24.1.

The vessel should be steamed for long enough to ensure the removal of residues – this may take in the order of hours,

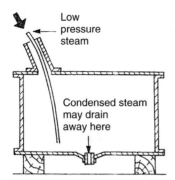

Low pressure steam

Condensed steam may drain away here

24.1   Steaming a vessel.

depending on the size and construction of the vessel and the nature of the residues being removed. There is a risk of electrostatic ignition during this process. Precautions will need to be taken to earth the vessel and the equipment.[173]

Small containers may be cleaned by water washing, using a solution of water and a detergent or caustic agent. This cannot be done effectively cold. The vessel should be completely immersed in the cleaning solution and boiled for at least 30 min. Jet washing may be used or proprietary equipment that combines jetting and steaming. Once cleaned, the vessel must be thoroughly inspected, to ensure that no residues remain.

Inerting

When the vessel cannot be cleaned or made safe by gas freeing, it may be worked on safely if the atmosphere inside the vessel is made inert by excluding oxygen. Some residue materials still present a danger when there is no oxygen in the atmosphere – these include nitrates and ethylene oxide. These must be removed.

The simplest and easiest way to inert a small vessel is to fill it completely with water, see Fig. 24.2. This is particularly useful in applications such as hot work in scrapyards, because there is no need for the use of a flammable gas detector to check the atmosphere.

Where the vessel is large, two points should be taken into consideration: will the vessel be able to withstand the load imposed by the weight of the water and will it be possible to dispose of the large quantity of contaminated water safely?

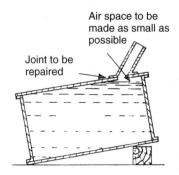

24.2   Welding over water.

Inerting may be carried out with gases such as carbon dioxide and nitrogen. Used on their own, there is some difficulty in ensuring that the whole vessel is uniformly filled, since this relies on diffusion. A slight positive pressure will need to be maintained, to prevent air entering.

Foams are used to inert vessels. These are visible and ensure positive displacement of the air inside the vessel. The most effective for work on vessels that have contained low flashpoint materials is nitrogen foam, which is produced using nitrogen, water and a detergent foam compound. After the hot work has finished, the foam eventually decays into a water mixture.

Confined Spaces

A confined space is defined by UK legislation[176] as anywhere which is more-or-less enclosed in which there is a foreseeable risk of fire or explosion or of being overcome by a flowing fluid or solid, or where the atmosphere may cause loss of consciousness due to overheating or exposure to gas, fume, etc. In the USA[177] a confined space is defined as a relatively small or restricted space, where there is a health or safety hazard. The accident record in confined spaces is very bad – more than 50% of those killed in them were attempting to rescue someone else.

Entry into confined spaces should be avoided by doing the job another way if it is possible. However, where entry is necessary, certain precautions should be taken beforehand. Permits-to-work will generally be required.

Confined spaces may present several increased risks:

- Asphyxiation from the shielding gases
- Lack of oxygen caused by chemical reaction (e.g. rusting) or poor ventilation
- Increased oxygen levels related to the welding equipment[178]
- Toxic vapours or fumes from the contents or the welding operation
- Increased risk of electric shock (see next section)
- Dusts
- Hot conditions
- Restricted access, e.g. entry via a manhole.

Mechanical and electrical isolation of the space is essential if equipment within the space could operate inadvertently. For example, if welds are to be carried out inside a mixing vessel, then the mixing machinery would need to be isolated. The vessel should be cleaned before entry if it contained anything that could give rise to fumes that are hazardous to health during welding operations. Forced ventilation may be necessary to maintain an adequate supply of air. Oxygen should never be used to sweeten the air. Portable gas cylinders should be left outside the space. Leaks of either fuel or oxygen into the space are potentially hazardous.

The air should be tested to ensure that it contains sufficient oxygen to support life[179] and that it is free from substances hazardous to health above their respective exposure limits. This should be done both before entry, but also while work is in progress. If the air inside the vessel cannot be guaranteed to be fit to breathe, or free from hazardous substances above their exposure limits, then respiratory protective equipment or breathing apparatus will be required. This was discussed in Chapter 6.

Half masks that fit over the nose and mouth are available for welders, see Fig. 24.3. This type of mask filters out particulates. It is essential that the respirator is a good fit over the face – beards could prevent a good fit being obtained. The filter must be of the correct type to remove fume, and it should be remembered that such a mask cannot remove gases. The respirator should be comfortable to wear, and must be compatible with the welder's other protective equipment.

Breathing apparatus with an independent supply of breathable air permits work in atmospheres that are immediately dangerous to life (but not explosive atmospheres). The air may be supplied from a line, in which case the air must be of breathable quality[82–85] and it

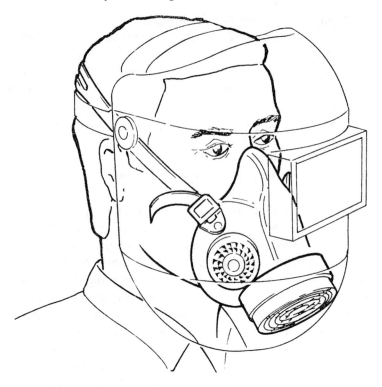

24.3   Dust respirator.

must not be possible to connect the line inadvertently to another gas source. Alternatively the air may be supplied from a cylinder, see Fig. 24.4. Operators must be trained to use such equipment, which must be maintained in good order. If the apparatus should fail, there should be an emergency plan in place to rescue the worker.

Any equipment that is taken into the confined space, e.g. welding equipment, should be readily turned off from outside the space in the event of an emergency. When work is finished, all equipment should be removed to avoid allowing gas to leak into the area.

Emergency arrangements should be planned before entry. This may include training and rehearsal of emergency shutdown and rescue arrangements.

Situations of Increased Risk of Electric Shock

When arc welding, the welder is holding apparatus that is in part live. By the use of safe equipment, such as insulated electrode

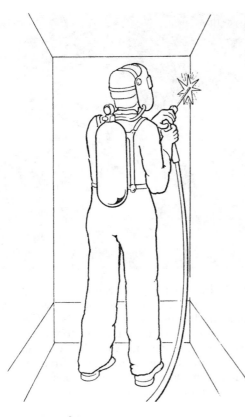

24.4   Breathing apparatus.

holders and dry insulating gloves the welder can normally avoid receiving an electric shock. However, the risk is always present. The open circuit voltage of most welding sets is sufficient to cause a fatal shock. This becomes more likely in certain environments, e.g. where:

— The welder is hot and sweaty.
— The environment is wet.
— The conditions are cramped.
— The environment is conducting (e.g. a steel vessel).

In these situations, extra precautions should be taken. Fatal accidents have occurred when welders have made a conducting path between live parts of the welding equipment and earthed metal work nearby. For example, a welder was killed when his head touched the steel vessel that he was working in.

24.5   Pipeline welding on a high pressure, 1 m diameter, gas
supply line. Welding power is supplied from DC generators on
the crawler units, minimising electric shock risks. The welder
furthest away is wearing a helmet with cape to prevent ingress
of spatter when working under the pipe. The overhead canopies,
with fume extractor hoses, provide protection from inclement
weather when required. Other safety precautions are not
forgotten – a marker for overhead electric cables is just visible
above.

The welding set

Welding sets should be purchased with a degree of ingress protection (IP) that is suited to the environment. European markings are
given in BS EN 60529[180] as the 'IP' code. Welding sets would normally be IP21 to IP23, where IP21 protects against dripping water,
and IP 23 protects against water spray up to 60° from the vertical.
In an outdoor application, the higher IP is essential, see Fig. 24.5.

The open circuit voltage can be reduced. Sets are available that
are manufactured for use in conductive environments and these are
marked with a letter S.[145] The open circuit voltage of these sets is
restricted so that it is no more than 48 V rms (root mean square) AC,
or 113 V DC peak.[145]

Voltage reduction devices are also available. These interrupt the
welding supply on open circuits and ensure that only 20–30 V
appears at the electrode. US standards recommend not greater than

38 V rms AC or 50 V DC.[147] The welder touches down the electrode and the device responds by restoring the welding supply to enable the arc to be struck. These must be designed to fail to safety. They should not be possible to by-pass and there should be an indication (e.g. a lamp) to show that they are working.

Electrode holders and all connections should be fully insulated. The welder should use insulating stands and mats, and adequate dry protective clothing to minimise the amount of bare skin – gloves, boots, overalls, etc.

Environments Containing Substances Hazardous to Health

While welding itself produces fume that may be hazardous to health, the welder sometimes works in an environment that is even more hazardous.

Work near asbestos

Repair welding may well bring a welder into an environment containing asbestos. The welders should be taught how to recognise asbestos and what to do. Asbestos should not be disturbed. A decision will need to be made regarding its removal. Any work in the vicinity of asbestos should be done in accordance with local regulations.[69,70]

Work on batteries

In lead-acid and other secondary batteries, the electrochemical decomposition of water forms an explosive mixture of hydrogen and oxygen – 'gassing', especially during charging. This must be removed before attempting to weld the lead or lead alloy terminals. Blowing out with low pressure compressed air must be done carefully to avoid spraying acid electrolyte. Note that no welding operations should be carried out where batteries are on charge or have recently been charged. Charging rooms should carry notices prohibiting naked lights. Welding of lead is liable to produce a significant quantity of fume, which will need to be controlled below the exposure limits.[67,68]

Work on coated materials

Where coated materials are to be welded, the fume will contain degradation products from the coating in addition to the parent metal. The coating should be assessed and removed if necessary.[74] Some coatings may be left and the requirement for local exhaust increased, for instance it is possible to weld galvanised steel with good local exhaust ventilation or on-gun extraction. In this particular application, both zinc and zinc oxide fume are captured. The finely divided zinc is pyrophoric. This can create a fire risk in the fume extraction system.

Part 3

Legislation and Appendices

25

Legislation

United Kingdom

Availability of documents

Her Majesty's Stationery Office has developed a website where the full text of the Statutory Instruments and Acts of Parliament may be viewed and printed. The legislation contained on the website is subject to Crown Copyright protection, but it may be reproduced free of charge provided that it is reproduced accurately and that the source and copyright status of the material is made evident to users. The address is *http://www.legislation.hmso.gov.uk/* . Currently, legislation dating back to 1987 is available. The legislation (of any age) is also available as printed documents from Her Majesty's Stationery Office.

Health and Safety at Work etc. Act 1974

This is an 'enabling' Act;[1] it states duties in general terms and enables specific regulations to be made. Employers and self-employed persons must ensure, as far as is reasonably practicable, the health, safety and welfare at work of all employees and the health and safety of other persons who are affected by their undertaking. In particular they must, as far as is reasonably practicable:

1 provide and maintain safe plant and systems of work;
2 arrange for the safe use, handling, storage and transport of articles and substances;
3 provide the necessary information, instruction, training and supervision;

4   maintain a safe place of work, and provide and maintain safe means of access and egress;
5   provide and maintain a safe working environment.

The designer, manufacturer, importer or supplier of any article used at work is required to ensure that it is designed and constructed so that it is safe when it is being set, used, cleaned or maintained. Adequate information is to be provided to ensure that persons know the use for which it was designed and any special conditions that are necessary to ensure safety. Similar duties apply to the supply of substances for use at work, and in particular there is a duty to supply information. Employees must take reasonable care for their safety and that of others, and cooperate with their employers.

Enforcement of the Act is via the Health and Safety Executive or the local authority, depending on the type of premises. Enforcement officers have wide-ranging powers of investigation, and powers to:

1   prosecute, if they find serious breaches of the statutory requirements;
2   issue Improvement Notices, requiring the improvement of conditions, which in the inspector's opinion contravene a specific statutory requirement, within a stated period;
3   issue Prohibition Notices, demanding that work shall cease, when the inspector is of the opinion that there is a risk of serious personal injury.

There is an appeal process for notices.

The Act is general and has now been interpreted in more detail by regulations made under the Act. Much health and safety law in the UK is goal-setting, specifying what must be achieved rather than the means to achieve it. There are approved codes of practice and a large number of guidance notes that interpret the law and give practical advice about how to comply with it.

The Factories Act, 1961 has not been repealed, but much of it has now been replaced by regulations made under the Health and Safety at Work, etc. Act.

Management of Health and Safety at Work Regulations, 1999
(Statutory Instrument no. 3242)

The Management of Health and Safety at Work Regulations, 1999[181], which are a revision of the 1992 regulations, require an employer to set up a management system so that health and safety arrangements are assured. This system will include assigning responsibilities, assessing risks and risk control measures, and implementing controls and training for the workforce.

Provision and Use of Work Equipment Regulations, 1998

These Regulations[7] require an employer to provide equipment that is suited to the work, with guarding as appropriate to prevent access to danger areas. The work equipment should have controls and means of isolation, and it should be maintained.

Fire Precautions Act 1971

This Act[30] requires certain premises to have a valid fire certificate. Such premises included those where more than 20 people are at work at any one time, or where explosive or highly flammable materials are stored or used. Thus most employers undertaking welding operations probably require a fire certificate. This specifies the minimum standards required of safe means of escape, means of fighting fire, warning systems and training requirements.

Those premises which do not require a fire certificate must still have provision for safe means of escape, means of fighting fire, warning systems and training. Fire hazards must be included in the employers' risk assessments, whether the premises have a fire certificate or not.

Reporting of Injuries, Diseases and Dangerous Occurrences Regulations 1995 (Statutory Instrument no. 3163)

These Regulations[28] require injuries, diseases and dangerous occurrences to be reported to the enforcing authority. The types of accident that must be reported include: fatality, fractured bones, amputation, electric shock, loss of consciousness due to asphyxia or exposure to a harmful substance, a chemical or hot metal burn to the eye. Diseases that are listed include some conditions related to the use of vibrating tools, poisonings by such substances as beryllium, cadmium, lead, manganese, oxides of nitrogen, and occupational asthma associated with the welding of stainless steel and use of rosin-based solder fluxes. Dangerous occurrences include collapse or overturning of lifts, hoists, cranes, fork-lift trucks, failure of pressure vessels, unintentional contact with overhead electric cables, malfunction of radiation generators and failure of radiation sources to be returned to their containers after use. The full list should be consulted and a guidance document is available.

Control of Lead at Work Regulations 1998
(Statutory Instrument no. 543)

These Regulations[67] require that exposure of employees to lead is prevented, or where this is not reasonably practicable, adequately controlled. The employer shall not embark on work with lead without first assessing whether exposure is liable to be significant. As far as is reasonably practicable, prevention or control of exposure shall be secured by measures other than personal protective equipment. However, where the risk of significant exposure remains, the employer shall provide the employee with suitable and sufficient protective clothing. Clothing provided for protection shall either comply with the Personal Protective Equipment Regulations 1992,[8] or be approved by the Health and Safety Executive.

An Approved Code of Practice gives practical guidance.[67] Lead burning, welding and cutting of lead coated and painted work, and dry grinding of lead are listed as types of work where there is liable to be a significant exposure to lead. Soldering and handling of clean, solid metallic lead are listed as those where there is not liable to be significant exposure. The occupational exposure limit for lead in the atmosphere is given as $0.15\,\mathrm{mg\,m^{-3}}$.

Health and Safety (Safety Signs and Signals) Regulations 1996
(Statutory Instrument no. 341)

A guide to these Regulations is given in a book.[182] These Regulations standardise the presentation of signs, both with respect to colour and shape, see Table 25.1.

Health and Safety (First Aid) Regulations 1981
(Statutory Instrument no. 917)

These Regulations[24] require the provision of appropriate equipment, facilities and trained persons for the workplace. Records must be kept and employees informed of the arrangements. Guidance on

Table 25.1. Standard colour and shape for safety signs (UK)

Safety colour	Shape	Purpose	Uses
Yellow	Triangular	Warning	Take care, examine, e.g. electrical hazard signs
Red	Round	Prohibition	Dangerous behaviour, e.g. no smoking signs
Blue	Round	Mandatory	Specific behaviour required, e.g. wear hearing protection
Green	Rectangular	Emergency escapes, first aid	Labels for doors, exits, facilities

these Regulations is in a book which suggests the numbers of first aiders required in different circumstances. There are nationally recognised training courses.

Ionising Radiations Regulations 1999
(Statutory Instrument no. 3232)

To embark on work with ionising radiation[103] now requires prior authorisation by, and notification of, the HSE. The employer must carry out a risk assessment in order to identify the measures that need to be taken to restrict the exposure of his or her employees. The basic aim is to prevent the occurrence of the acute effects of radiation by keeping the doses below the threshold level, and to limit the probability of the longer term effects to a level that is as low as is reasonably achievable. Engineering controls and design features are required as the first priority, administrative controls are also needed and appropriate personal protective equipment must be provided where adequate control is not achieved by other means.

Any person liable to be exposed to an annual effective dose in excess of 6 mSv, or an equivalent dose greater than three tenths of a relevant dose limit, shall be made a classified worker, and subject to dose assessment (dosimetry) and medical surveillance. Further information is in the Approved Code of Practice and Guidance.[103]

Control of Substances Hazardous to Health Regulations 1999
(Statutory Instrument no. 437)

These Regulations[76] apply to any substance that can be hazardous, but for the welder the main interest is in substances that can be inhaled. Substances that are covered by these Regulations are:

1   those for which there is a maximum exposure limit or an occupational exposure standard;
2   those that are classified as very toxic, toxic, harmful, corrosive or irritant;
3   a biological agent;
4   any dust present in air at a concentration greater than $10\,mg\,m^{-3}$ total inhalable dust, or $5\,mg\,m^{-3}$ respirable dust;
5   any other substance that creates a hazard comparable to the above.

Employers must assess the risks to health created by their work and prevent or control exposure to substances hazardous to health. Exposure must be prevented where reasonably practicable, but where not, the employers should consider enclosure, local exhaust ventilation and, as a last resort, personal protective equipment. Any control measures must be properly used and maintained. In certain circumstances monitoring exposure and health surveillance are appropriate. Employees shall be given adequate information, instruction and training. Further information is available in an Approved Code of Practice.[76]

Electricity at Work Regulations 1989
(Statutory Instrument no. 635)

These Regulations[18] apply to uses of electricity at work from all sources. There are no voltage limits in the Regulations, the criterion for action being the presence of danger or potential to harm persons. They require systems to be of adequate construction and electrical equipment to be of construction suited to the environment where they are to be used. This would include the provision of special equipment for work out of doors or in an explosive atmosphere.

Conductors that give rise to danger shall be insulated or protected and placed so as to prevent danger. This would apply in a workshop, where trailing cables must be protected from damage by fork lift trucks, abrasion by sharp pieces of metal, etc. All connections must be mechanically and electrically sound. Earthing or other suitable means is required to prevent a conductor from becoming charged as a result of a fault.

There must be a means for cutting off the supply which is readily accessible, and the systems must have protection from foreseeable excess currents due to faults, overloads, etc. Further guidance is available.[18]

Noise at Work Regulations 1989
(Statutory Instrument no. 1790)

These Regulations[105] require employers to make an assessment where employees are liable to be exposed to a daily noise exposure of 85 dB(A) or above. Noise is to be reduced at source where reasonably practicable, or other engineering measures introduced. Where workers are exposed to between 85 dB(A) and 90 dB(A) daily exposure, they must be provided with suitable hearing protection on request. If the exposure exceeds 90 dB(A) then hearing protection must be provided and the wearing of it is mandatory. Employees must be given information, instruction and training, so that they know the risks and how to protect themselves. There is a guidance document.[105]

Pressure Equipment Regulations 1999[183] (See below)
(Statutory Instrument no. 2001)

Pressure Systems Safety Regulations 2000
(Statutory Instrument no. 128)

These Regulations[183,184] require that pressure vessels are designed and constructed to follow good engineering practices. They describe the tests that must be done. The 2000 Regulations describe the schemes of examination and testing that need to be done on certain pressure systems.

Highly Flammable Liquids and Liquefied Petroleum Gases Regulations 1972 (Statutory Instrument no. 917)

These Regulations[185] apply to propane and LPG, and to highly flammable and extremely flammable liquids. They relate to the safe storage, marking of containers, and fire precautions.

Control of Asbestos at Work Regulations 1987, as amended69

Before any work with asbestos is started the employer must make a thorough assessment of the likely exposure. The employer must make a suitable plan of work before removing any asbestos, and most work must be notified to the enforcement authority. Employees liable to be exposed must be provided with adequate information, instruction and training to understand the risks and the necessary precautions that must be taken. Exposure must be reduced to the lowest level reasonably practicable – areas where asbestos work is being carried out will be clearly marked and demarcated, kept clean, and monitored. Special clothing will be required and all control measures must be kept in clean efficient working order. The Regulations provide for health records to be maintained. Asbestos waste may only be disposed of in the approved manner.

Personal Protective Equipment at Work Regulations, 1992

These Regulations[8] require that appropriate personal protective equipment is provided, such as gloves, goggles, face shields, hearing protections, etc, where the risk has not been adequately controlled. It must be viewed as the last resort, with engineering control being preferred. The employer must choose equipment as a result of risk assessment, and the equipment must fit, be effective, and be maintained or replaced as necessary. Employees must be trained to be able to make effective use of such equipment.

'New Approach' Directives

These Directives, which emanate from the European Union, set out essential requirements that must be met before an item can be placed on the market in Europe. Once these requirements have been met, then the article can be 'CE' marked, and marketed. Employers should look for the CE mark on the items that they buy, but they should be aware that it is not a guarantee of quality – it merely means that the manufacturer claims conformity. Directives of interest to health and safety include the Personal Protective Equipment Directive 89/686/EEC and the Machinery Safety Directive 89/392/EEC and 91/368/EEC. Directives from Europe are implemented in the UK by means of Regulations.

United States of America

Availability of Documents

The OSHA Act and all Codes of Federal Regulations mentioned in this book are available from the US Government website. They may be found by following links from *http://www.osha.gov/* . The Regulations themselves have links to compliance letters and interpretations.

Occupational Safety and Health Act, 1970

This Act[5] had the stated aim 'to assure safe and healthful working conditions for working men and women'.

The general requirement is stated in section 5 of the Act, which states that each employer shall furnish for each of their employees employment and a place of employment which are free from recognised hazards that are causing or are likely to cause death or serious physical harm to their employees. Employers shall comply with the occupational safety and health standards promulgated under the act. Each employee shall comply with the occupational safety and health standards and all rules, regulations and orders issued pursuant to the act which are applicable to their own actions and conduct.

The detail has been fleshed out in various parts of the Code of Federal Regulations. Many are quite old and there are variations from state to state, so the employer needs to check what the detailed local requirements are. For welding there is a Standard, Z49.1 1999[124] which is much more up-to-date than the Federal Regulations. OSHA encourages employers to abide by the more current industry consensus standards because they are more likely to be abreast of the state of the art than the applicable OSHA Standard may be.

29 CFR 1904 Recording and reporting occupational injuries and illnesses

These regulations[29] require employers to keep records and make reports of occupational accidents and illnesses, to aid in the development of information regarding their causes and prevention. The records are in the form of a log, 'OSHA No. 200' (soon to be replaced by no. 300).

The categories of recordable incidents include fatalities, cases resulting in lost workdays or transfer to another job (or termination of employment). They include incidents requiring medical treatment (other than first aid) or involving loss of consciousness. They also include diagnosed occupational illnesses.

29 CFR 1910 Occupational safety and health standards

This Standard contains many subparts of specific importance to the welding environment.

The 29 CFR 1910. etc are all Occupational Health and Safety Standards. They can be accessed from the website by following the links.[187] The 29 CFR 1910 (and others) are grouped together into subparts which contain sets of standards related to the same topic, so sometimes it is reasonable to refer (the reader) to the whole set. At other times, the reference is to the particular standard within the group. Thus, for example, subpart F to 29 CFR 1926 contains the Standards related to the Construction Industry, and within this subpart 29 CFR 1926.352 is the specific standard for fire protection.

– **Subpart E, 1910.35–38 Means of egress**
  This subpart describes the emergency means of escape that must be provided and specifies the elements that must go into any emergency plan.

– **Subpart G, 1910.94–98 Occupational health and environmental control[107]**
  This subpart contains the requirements for ventilation, noise and non-ionising radiation.

– **Subpart H, 1910.101–126 Hazardous materials**
  This deals with compressed gases: acetylene, hydrogen and oxygen, and the storage of highly flammable liquids.

– **Subpart I, 1910.132–139 Personal protective equipment**
  This subpart contains requirements for eye, face, head, foot, hand and respiratory protection. It also contains requirements for electrical protective devices (blankets, gloves, etc).

– **Subpart J, 1910.141–147 General environmental controls[14]**
  This contains details of sanitation, colour codes for marking physical hazards and accident prevention signs. It also contains the requirements for entry to confined spaces, and lockout and tagout procedures.

– **Subpart K, 1910 Medical and first aid**
  This has the requirements for first aid, in part 1910.151.

– **Subpart L, 1910.155–165 Fire protection**[33]
This contains the requirements for means of fighting fire, detecting fire and giving warning.

– **Subpart O, 1910.211–219 Machinery and machine guarding**[119,162]
In particular this subpart has regulations dealing with abrasive wheel machinery, and forging machines. It also contains the general requirements for all machines.

– **Subpart P, 1910.241–244 Hand and portable powered tools and other hand-held equipment**
This subpart requires the employer to take responsibility for the safe condition of the tools and equipment. It specifies the guarding requirements.

– **Subpart Q, 1910.251–255 Welding, cutting and brazing**[34,124]
This subpart contains regulations dealing with oxygen fuel gas welding and cutting, arc welding and cutting and resistance welding. It deals with the workplace, the equipment to be used and the work practices.

– **Subpart S, 1910.301–399 Electrical – general**[20]
This subpart contains standards relating to wiring design, wiring methods and hazardous locations. It also describes the training requirements, the selection and use of work practices and the safeguards for personnel protection.

– **Subpart Z, 1910.1000–1450 Toxic and hazardous substances**[42,72]
This subpart contains the requirements for restriction of exposure to substances hazardous to health, including tables with the exposure limits.
   There are separate sections dealing with the following:
   • 1910.1001 Asbestos[70]
   • 1910.1018 Arsenic[71]
   • 1910.1025 Lead[68]
   • 1910.1027 Cadmium[170]
   • 1910.1096 Ionizing radiation.[104]

29 CFR 1915 Occupational safety and health standards for shipyard employment37,126

These regulations contain:

– Subpart B: confined and enclosed spaces;
– Subpart D: welding, cutting and heating;
– Subpart H: tools and related equipment;

– Subpart I: personal protective equipment;
– Subpart L: electrical machinery;
– Subpart Z: toxic and hazardous substances.

29 CFR 1926 Safety and health regulations for construction[36,74,108,127]

These regulations contain:

– Subpart I: tools – hand and power;
– Subpart J: welding and cutting;
– Subpart K: electrical;
– Subpart Z: toxic and hazardous substances.

Appendix A: Glossary of Terms

Backfire: flame going back into the blowpipe neck or body. It may be self-extinguishing or may be sustained. This results in a 'popping' or 'squealing' noise, with a small pointed flame, or as a rapid series of small explosions.

Bar: a measurement of pressure – approximately one atmosphere.

Braze welding: joining pieces of metal using a filler metal with a lower melting point than that of the sections to be joined, without capillary action.

Brazing: making a joint between pieces of metal in which molten filler metal is drawn by capillary attraction into the space between the closely adjacent surfaces of the parts to be joined. In general, the melting point of the filler is above approximately 450 °C and below that of the metal to be joined.

Confined space: a space in which there is a risk of asphyxiation or explosion, due to its enclosed nature and poor ventilation.

Cut-off valve, pressure sensitive: a valve that automatically closes when there is a sudden back pressure from the downstream side of the valve.

Cut-off valve, temperature sensitive: a valve that automatically closes off the gas supply before a flame arrestor reaches a sufficiently high temperature to ignite the gas on the upstream side.

Flame arrestor or flashback arrestor: a device that arrests a flame, travelling in either direction.

Flashback: the flame travels back beyond the blowpipe body into the hose; it is potentially serious.

Fume: solid particles in the air that have originated from metal and fluxes that have been heated in the weld.

Guard: physical barrier, preventing access to the dangerous parts of the machine. A guard must be sturdily built and require a tool to remove it.

Interlock: a safety switch on an opening that ensures that equipment is shut down if the door is opened. An interlock should not reset itself if the door is closed; resetting should be a positive action.

Ionising radiation: X-rays, alpha particles and gamma rays. Capable of giving rise to chemical changes in the body.

Laser: light amplification by stimulated emission of radiation. This produces a light beam of a single wavelength, of small divergence and high power.

Non-ionising radiation: heat, light and ultraviolet radiation.

Non-return valve: a self-actuating valve that prevents gas from flowing in the reverse direction.

Pressure gauge: a device that indicates pressure by means of a needle rotating over a calibrated dial.

Pressure regulator: a device for delivering a constant outlet pressure, from a cylinder of varying inlet pressure.

Snifting: the rapid opening and closing of the main valve on a gas cylinder to remove dust, etc. This should never be done on a hydrogen cylinder.

Soldering: a joining process similar to brazing, but in which the filler metal melts at a temperature below approximately 450 °C.

Spatter: droplets of liquid metal that are ejected from the weld pool.

Surfacing: depositing a layer of a metal on a substrate of the same or different metal by a process involving heat.

Thermal cutting: parting or shaping metal by the application of heat with or without a stream of cutting oxygen or inert gas.

Thermal spraying: spraying material which had been melted in a spray gun, in a finely divided form, projected on to a suitably prepared substrate by a gas stream.

Welding: uniting pieces of metal, the 'parent metal', at joint faces melted by heat; in many cases additional 'filler metal' of similar composition to the parent metal may be added in making the joint. Also applied to thermoplastics materials, but in welding these, their surface is only heated to their plastic range.

Appendix B: Risk Assessment for Arc Welding

There are many ways of undertaking risk assessment, from the purely qualitative to the quantitative. All methods have their advantages and disadvantages, depending on the type of work.

The following is a qualitative risk assessment for welding mild steel in dry conditions, using manual metal arc (MMA).

It has the essential features, identifying:

- the hazards;
- the significant risks within the workplace;
- who is affected;
- what the existing control measures are;
- what standard is to be achieved – the legal requirements being the lowest level acceptable;
- an action plan.

The risk assessment must be communicated to those who are affected. It should be reviewed at intervals and whenever there is a significant change in risk or an accident may show that it is not valid.

While the attached is a sample risk assessment, each employer must do his own, since each workplace is different. It is not claimed that the following is exhaustive.

Table A.1. Sample risk assessment

Location of the work ....

Description of the work .... MMA welding of mild steel in a dry environment

The hazard	How harm arises	The effect	Who is affected	Existing control measures	Standards that should be achieved
Electricity	Contact with mains supply, due to faulty or degraded wiring	Electric shock, possibly fatal	Welder, or anyone who touches equipment	(The employer should fill this column in)	Equipment should be built to a standard, installed by a competent electrician. Cables, etc, should be visually inspected to check that they have not been damaged. The means to switch off the equipment should always be readily accessible and easily identified
	Internal faults in equipment	Electric shock, possibly fatal	Anyone who touches the equipment		Equipment that is class I should be earthed. The workpiece should be fitted with an earth lead if the welding set does not have reinforced insulation. Portable and transportable equipment should be formally tested at prescribed intervals. Personnel should be encouraged to report faults or accidental damage to equipment to a nominated person
	Stray currents damaging protective earths, chains, slings, motors, etc	Damaged equipment. Possible loss of protective earth in affected equipment – electric shock	Anyone who touches the equipment		Welders should be trained to make good connections by making contact with bright metal. Welding leads should be insulated, robust, and the connectors should not allow inadvertent contact

Table A.1. (continued)

The hazard	How harm arises	The effect	Who is affected	Existing control measures	Standards that should be achieved
	Contact with the OCV of the set	Electric shock, possibly fatal	The welder		Welders should be trained to adopt a safe procedure when changing electrodes. Electrode holders should be fully insulated
	Contact between welders working on different phases	Electric shock, possibly fatal, due to the increased voltage between them	The welders		Welders on different phases should be segregated, either by partitions or by distance
Fume and gases	Inhalation of fume and gases during welding	Irritation	The welder, and to a lesser extent others in the room		The manufacturer's safety data sheet (MSDS) for the electrode should be checked, and the composition of the parent material should be checked too. The fume must be controlled to a level that ensures that the welder has less than $5\,\text{mgm}^{-3}$ in his or her breathing zone. Local exhaust ventilation will probably be needed, in addition to good general ventilation. Any coatings on the steel will need to be assessed and removed if necessary (e.g. if containing cadmium). Any exhaust ventilation equipment must be maintained, and the welder taught how to use it properly
	Vapours from degreasing, etc	Irritation of respiratory tract, hazardous products if it reaches the welding arc	Anyone in the area		MSDSs must be consulted before use. Advice on engineering or other control measures should be taken

Hazard	Cause	Harm	Who is at risk	Control action
Radiation	Light from the arc entering the eye or falling on the skin	Arc eye, burns to the skin	Welders and anyone close to the arc	Welders and their helpers should wear clothing to protect all their skin. Welders and others close to the arc should wear visors with filters to cut out the harmful radiation (to the British Standard). People at greater distances should be protected by the use of welding curtains or screens
Fire	Sparks and spatter from the weld falling on combustible material	Loss of life and property	Everyone	The area should be prepared for welding by removing all combustible material or covering it with fireproof blankets or shields. The employer should have an emergency plan to ensure that all occupants can leave the building and go to a place of safety. Fire extinguishers of the appropriate type must be available. Personnel should be trained in the emergency plan and the operation of the fire fighting equipment
Noise and vibration	From grinding and deslagging operations	Loss of hearing, long term ill health due to vibration	The welders	Noise reduction should take priority. If noise levels exceed 90dB(A) for an 8hr day, hearing protection is mandatory. Between 85 and 90 hearing protection must be made available on request. The employer should choose tools that have low vibration where reasonably practicable. Where vibration levels are high, exposure can be reduced by reducing the time of use and the effects can be reduced by keeping warm

Table A.1. (continued)

The hazard	How harm arises	The effect	Who is affected	Existing control measures	Standards that should be achieved
Mechanical hazards	Falling objects	Injuries, especially to feet	Welders and helpers		Welders and helpers should wear safety footwear. Only those trained to use cranes should be allowed to use them. Lifting equipment must be maintained and inspected
	Slag and other particles entering the eye	Serious eye injury	Welders and helpers		Welders and others should wear goggles or a face visor when grinding or deslagging (BS EN 166 class B)
	Slips, trips and falls	Fractures, etc	Anyone using the workshop		Maintain an adequate level of housekeeping. Maintain an adequate level of lighting
Manual handling	Lifting tasks	Back injuries	Anyone undertaking the work		Training is to be given in correct manual handling technique. Manual handling tasks are to be assessed before they are carried out. Any heavy loads should be lifted using aids where possible

Action Plan, (to make up the differences between the existing control measures and the standards that should be achieved).

Assessment carried out by ....   Date ....   Review date ....

Appendix C: Useful Addresses and Abbreviations

ANSI
American National Standards Institute
11 West 42nd Street, 13th Floor, New York, NY 10036-8002, USA

AWS
American Welding Society
550 NW Lejeune Road, Miami, Florida 33126, USA

BCGA
British Compressed Gases Association
14 Tollgate, Eastleigh, Hampshire, SO53 3TG, UK

BSI
British Standards Institution
389 Chiswick High Road, London, W4 4AL, UK

CGA
Compressed Gases Association
1725 Jefferson Davis Highway, Suite 1004, Arlington, Virginia 22202, USA

HMSO
Her Majesty's Stationery Office Ltd, Customer Services, St Crispin's, Duke Street, Norwich, NR3 1GN, UK

HSE
Health and Safety Executive, HSE Books
PO Box 1999, Sudbury, Suffolk, CO10 2WA, UK

IOSH
Institution of Occupational Safety and Health
The Grange, Highfield Drive, Wigston, Leicestershire, LE18 1NN, UK

Laser Institute of America
12424 Research Parkway, Orlando, Florida 32826, USA

NEMA
National Electrical Manufacturer's Association
1300 North 17th Street, Suite 1847, Rosslyn, Virginia 22209, USA

NFPA
National Fire Protection Association
Customer Service Department, 1 Batterymarch Park, Quincy, Maryland 02269-9101, USA

NIOSH
National Institute of Occupational Safety and Health
4676 Columbia Parkway, Cincinnati, Ohio 45226, USA

NRPB
National Radiological Protection Board
Chilton, Didcot, Oxfordshire, OX11 0RQ, UK

OSHA
Occupational Health and Safety Administration
US Department of Labor, Occupational Safety and Health Administration, 200 Constitution Avenue, N.W., Washington DC 20210, USA

RIA
Robotic Industries Association
900 Victors Way, PO Box 3724, Ann Arbor, Michigan 48106, USA

RMA
Rubber Manufacturer's Association
1400 K Street NW, Suite 900, Washington DC 20005, USA

TSSEA
Thermal Spray and Surface Engineering Society
18 Hammerton Way, Wellesbourne, Warwickshire, CV35 9NT, UK

TWI
The Welding Institute
Granta Park, Great Abington, Cambridge, CB1 6AL, UK

US Government Printing Office
Congressional Legislative and Public Affairs, Government Printing Office, 732 North Capitol Street NW, Washington DC20401, USA

Woodhead Publishing Limited
Abington Hall, Abington, Cambridge, CB1 6AH, UK

References

For abbreviations, see Appendix C.

1 Health and Safety at Work, etc, Act 1974.
2 *Management of health and safety at work.* Management of Health and Safety at Work Regulations 1999. Approved Code of Practice L21, HSE Books, revised 2000.
3 Her Majesty's Stationery Office website: http://www.legislation.hmso.gov.uk/ . See Appendix C for postal address.
4 Health and Safety Executive website: http://www.hse.gov.uk/hsehome.htm . See Appendix C for postal address.
5 Occupational Safety and Health Act 1970.
6 Occupational Safety and Health Administration website: http://www.osha.gov/ .
7 *Safe use of work equipment.* Provision and Use of Work Equipment Regulations 1998. Approved Code of Practice and Guidance L22, HSE Books.
8 *Personal protective equipment at work.* Personal Protective Equipment at Work Regulations 1992. Guidance on the Regulations L25, HSE Books.
9 Health and Safety Law poster, ISBN 0 7176 2493 5.
10 'Health and Safety Law: what you should know', HSE Books, 1999, also available free of charge from www.hse.gov.uk/pubns/law.pdf .
11 OSHA poster 3165, www.osha-slc.gov/Publications/osha3165.pdf .
12 *Workplace health, safety and welfare.* Workplace (Health, Safety and Welfare) Regulations 1992, Approved Code of Practice and Guidance L24, HSE Books, revised 1996.
13 Walking-working surfaces, 29 CFR 1910 subpart D.
14 General environmental controls, 29 CFR 1910 subpart J.
15 'Lighting at work', HS(G) 38, HSE Books, 1998.
16 BS 8206: Part 1: 1985 Lighting for buildings, Code of Practice for artificial lighting, BSI.
17 *Manual handling.* Manual Handling Operations Regulations 1992, Guidance on the Regulations L23, HSE Books, 1998.

18 Memorandum of Guidance on the Electricity at Work Regulations 1989, Guidance on the Regulations HS(R) 25, HSE Books, revised 1998.

19 BS 7671: 1992 Requirements for electrical installations, IEE wiring regulations, 16th edition, BSI.

20 Electrical-general, 29 CFR 1910 Subpart S.

21 'Electricity at work, safe working practices', HS(G) 85, HSE Books, 1993.

22 'Maintaining portable and transportable electrical equipment', HS(G) 107, HSE Books, 1994.

23 International Commission on Non-Ionizing Radiation Protection, 'Guidelines for limiting exposure to time-varying electric, magnetic and electromagnetic fields (up to 300 GHz)', *Health Physics* 1998, **74** (4), 494–522.

24 *First aid at work*. The Health and Safety (First Aid) Regulations 1981. L74, HSE Books, revised 1997.

25 Medical services and first aid, 29 CFR 1910.151.

26 US Department of Labor, 'How to prepare for workplace emergencies', OSHA 3088, 1995.

27 'First Aid Manual', Dorling Kindersley, London, 1999 also published in the USA, and the authorised manuals of the other voluntary aid societies.

28 'A Guide to the Reporting of Injuries, Diseases and Dangerous Occurrences Regulations 1995' L73, HSE Books, revised 1999.

29 Recording and reporting occupational injuries and illness, 29 CFR 1904.

30 Fire Precautions Act 1971.

31 Fire Precautions (Workplace) Regulations 1997, as amended.

32 Home Office, Scottish Executive, Department of the Environment (Northern Ireland) and HSE, 'Fire safety: an employer's guide', HSE Books.

33 Fire protection, 29 CFR 1910 subpart L.

34 General requirements, welding, cutting and brazing, 29 CFR 1910.252.

35 Fire protection and prevention, 29 CFR 1926 subpart F (Construction industry).

36 Fire protection (welding and cutting), 29 CFR 1926.352 (Construction industry).

37 Fire prevention, welding, cutting and heating, 29 CFR 1915.52 (Shipyards).

38 NFPA 51B: 1999, Standard for fire prevention in the use of cutting and welding processes, NFPA.

39 BS EN 3, Portable fire extinguishers, BSI, 1996.

40 NFPA 10: 1998 Portable fire extinguishers, NFPA.

41 BS EN 1089: Part 3: Identification of contents of industrial gas containers, BSI, 1997.

42 Hazard communication, 29 CFR 1910.1200.

43 GN2 Guidance for the storage of transportable gas cylinders for industrial use, rev 2 1997, BCGA.

44 CP31 Safe storage and use of cylinders in mobile workshops and service vehicles, BCGA.

45 CHIS 4 Use of LPG in small bulk tanks, HSE Books, 1999.

46 CHIS 5 Small scale use of LPG in cylinders, HSE Books, 1999.

47 NFPA 55: 1998 Storage, use and handling of compressed and liquefied gases in portable cylinders, NFPA.

48 NFPA 58: 1998 Liquefied petroleum gas code, NFPA.

49 BS EN 60079: Part 14: Electrical apparatus for explosive gas atmospheres: electrical installations for hazardous areas (other than mines), BSI, 1997.

50 CP21 Bulk liquid argon or nitrogen storage at users' premises, rev 1, 1998, BCGA.

51 CP19 Bulk liquid oxygen storage at users' premises, rev 2 1996, BCGA.

52 NFPA 50: Bulk oxygen systems at consumer sites, NFPA, 2001.

53 CP6 The safe distribution of acetylene in the pressure range 0–1.5 bar, BCGA, 1998.

54 CP5 The design and construction of manifolds using acetylene gas from 1.5 bar to a maximum working pressure of 25 bar, rev 1 1998, BCGA.

55 NFPA 51: 1997 Standard for the design and installation of oxygen–fuel gas systems for welding, cutting and allied processes, NFPA.

56 BS EN 29539: 1992 Specification for material for equipment used in gas welding, cutting and allied processes, BSI.

57 GN3 Application of the manual handling operations regulations to gas cylinders, BCGA.

58 BS EN 345: Part 1: 1993 Safety footwear for professional use, specification, BSI.

59 BS EN ISO 2503: 1998 Gas welding equipment. Pressure regulators for gas cylinders used in welding, cutting and allied processes up to 300 bar, BSI.

60 GN7 The safe use of individual portable or mobile cylinder gas supply equipment, BCGA.

61 E4 Standard for gas regulators for welding and cutting, CGA, 1994.

62 E3 Pipeline regulators inlet connection standards (inlet connectors on removable pipeline regulators in welding and cutting), CGA, 1991.

63 'Medical aspects of occupational asthma', MS25, HSE Books, 1998.

64 *Preventing asthma at work.* 'How to control respiratory sensitisers', L55, HSE Books, 1994.

65 'Occupational exposure limits' EH40 (re-issued annually), HSE Books.

66 Limits for air contaminants Table Z1; and Toxic and hazardous substances Table Z2. 29 CFR 1910.1000.

67 *Control of lead at work.* Control of Lead at Work Regulations 1998, HSC, COP 2, HSE Books.

68 Lead, 29 CFR 1910.1025.

69 *The control of asbestos at work.* Control of Asbestos at Work Regulations 1987, Approved Code of Practice L27, HSE Books, revised 1999.

70 Asbestos, 29 CFR 1910.1001.

71 Arsenic 29 CFR 1910.1018.

72 Air contaminants: toxic and hazardous substances, 29 CFR 1910.1000.

73 'Assessment of exposure to fume from welding and allied processes', EH 54, HSE Books, 1990.

74 Welding in way of preservative coatings, 29 CFR 1926.354 (Construction industry).

75 J Moreton and N A R Falla 'Analysis of airborne pollutants in working atmospheres: the welding and surface coatings industries'. *Analytical Sciences Monograph No. 7*, The Chemical Society, London, 1980.

76 *General COSHH ACOP, carcinogens ACOP, and biological agents ACOP*, Control of Substances Hazardous to Health Regulations 1999, L5, HSE Books.

77 'The control of exposure to fume from welding, brazing and similar processes', EH 55, HSE Books, 1990.

78 F3.1-89 Guide for welding fume control, AWS, 1989.

79 N Jenkins (editor), 'The facts about fume', second edition, The Welding Institute (now known as TWI) 1986, available from Woodhead Publishing.

80 Z 9.2-1979. Fundamentals governing the design and operation of local exhaust systems, AWS.

81 'An introduction to local exhaust ventilation', HS(G) 37, HSE Books, 1993.

82 BS 4275: 1997 Guide to implementing an effective respiratory protective device programme, BSI.

83 G7.1 Commodity specification for air, CGA, 1997.

84 BS EN 141: 1991 Specification for gas filters and combined filters used in respiratory protective equipment, BSI.

85 'Respiratory protective equipment, a practical guide for employers', HS(G) 53, HSE Books, 1998.

86 BS EN ISO 10882: Part 1: 2001 Health and safety in welding and applied processes. Sampling of airborne particles and gases in the operator's breathing zone Sampling of airborne particles, BSI.

87 BS EN ISO 10882: Part 2: 2000 Health and safety in welding and applied processes. Sampling of airborne particles and gases in the operator's breathing zone Sampling of gases, BSI.

88 F1.5-96 Methods for sampling and analysing gases from welding and allied processes, AWS, 1996.

89 F1.1 1999 Method for sampling airborne particulates generated by welding and allied processes, AWS.

90 F1.4-97 Methods for analysis of airborne particulate generated by welding and allied processes, AWS, 1997.

91 F1.3 1999 A sampling strategy guide for evaluating contaminants in the welding environment, AWS.

92 F1.2-1999 Laboratory method for measuring fume generation rates and total fume emission of welding and allied processes, AWS.

93 ULR Ultraviolet reflectance of paint, AWS, 1976.

94 BS EN 470: Part 1: 1995 Protective clothing for use in welding and allied processes. General requirements, BSI.

95 Pr EN 12477 Standard for welders gloves, BSI.

96   BS EN 1598: 1998 Health and safety in welding and allied processes. Transparent curtains, strips and screens for arc welding processes, BSI.

97   NFPA 701 Fire tests for flame propagation of textiles and films, NFPA, 1999.

98   BS EN 169: 1992 Specification for filters for personal eye-protection equipment used in welding and similar operations, BSI.

99   BS EN 166: 1996 Personal eye protection – specifications, BSI.

100 Z87.1 Practice for occupational and educational eye and face protection, ANSI.

101 F2.2 Lens shade selector, AWS, 1997.

102 BS EN 379: 1994 Specification for filters with switchable or dual luminous transmittance for personal eye protectors used in welding and similar operations, BSI.

103 *Work with ionising radiation.* Ionising Radiations Regulations 1999. HSE Books, revised document L121, 2000.

104 Ionizing radiation 29 CFR 1910.1096.

105 *Reducing noise at work.* Guidance on the Noise at Work Regulations 1989. L108, HSE Books, revised 1998.

106 *Sound solutions.* 'Techniques to reduce noise at work', HS(G) 138, HSE Books, 1995.

107 Occupational noise exposure 29 CFR 1910.95.

108 Occupational noise exposure 29 CFR 1926.52 (Construction industry).

109 BS EN 352: 1993 Hearing protectors safety requirements and testing. Ear muffs, BSI.

110 BS EN 458: 1994 Hearing protection. Recommendations for selection, use, care and maintenance; Guidance document, BSI.

111 Musculo-skeletal disorders 29 CFR 1910.900 in draft at January 2001.

112 'Hand-arm vibration' HS(G) 88, HSE Books, 1994.

113 BS 6842: 1987 Guide to measurement and evaluation of human exposure to vibration transmission to the hand, BSI.

114 BS EN 28662: Part 1: 1993 Hand held portable power tools. Measurement of vibrations at the handle. General, BSI.

115 BS EN 28662: Part 2: 1995 Hand held portable power tools. Measurement of vibrations at the handle. Chipping hammers and riveting hammers, BSI.

116 *Safe use of lifting equipment.* Lifting Operations and Lifting Equipment Regulations, 1998. Approved Code of Practice and Guidance L113, HSE Books.

117 NFPA 79 Electrical standard for industrial machinery, NFPA, 1997.

118 'Safety in the use of abrasive wheels', HS(G) 17, HSE Books, 2000.

119 Abrasive wheel machinery, 29 CFR 1910.215.

120 'Industrial robot safety' HS(G) 43, HSE Books, 2000.

121 R15.06: 1999 Industrial robots and robot systems – safety requirements, RIA.

122 EW-8: 1994 Recommendations for components of robotic and automatic welding installations, NEMA, also document D16.2, AWS.

123 'The safe use of compressed gases in welding, flame cutting and allied processes', HS(G) 139, HSE Books, 1997.

124   Z49.1: 1999 Safety in welding, cutting and allied processes, AWS.

125   Oxygen fuel gas welding and cutting 29 CFR 1910.253.

126   Gas welding and cutting 29 CFR 1915.55 (Shipyard employment).

127   Gas welding and cutting 29 CFR 1926.350 (Construction).

128   CP7 The safe use of oxyfuel gas equipment (individual portable or mobile cylinder supply), BCGA, 1996.

129   BS EN 559: 1994 Gas welding equipment: rubber hoses for welding, cutting and allied processes, BSI.

130   IP7 Specification for rubber welding hose, RMA, 1999.

131   SB-11 Use of rubber welding hose, CGA, 1996.

132   BS EN 1256: 1996 Gas welding equipment, specification for hose assemblies for equipment for welding, cutting and allied processes, BSI.

133   BS EN 560: 1995 Gas welding equipment, hose connections for welding, cutting and allied processes, BSI.

134   E-1 Standard connections for regulator outlets, torches and fitted hose for welding and cutting equipment, CGA, 2000.

135   BS EN ISO 5172: 1997 Manual blowpipes for welding, cutting and heating. Specification and tests, BSI.

136   CP 17 The repair of hand-held blowpipes & gas regulators used with compressed gases for welding, cutting & related processes. Revision 1: 1998, BCGA.

137   BS EN 730: 1995 Gas welding equipment. Equipment used in gas welding, cutting and allied processes. Safety devices for fuel gases and oxygen or compressed air. General specification, requirements and tests, BSI.

138   TB-3 Hose line flashback arrestors, CGA, 1998.

139   C4.2 Operator manual for oxyfuel gas cutting, AWS, 1990.

140   SB-8 Use of oxyfuel gas welding and cutting apparatus, CGA, 1998.

141   SB-4 Handling acetylene cylinders in fires, CGA, 1997.

142   'Health and safety in arc welding', HS(G) 204, HSE Books, 2000.

143   T Lyon et al., 'Evaluation of the potential hazards from actinic ultraviolet radiation generated by electric welding and cutting arcs', US Army Development Hygiene Agency, AD/A-033 768. Available from Microinfo Limited, PO Box 3, Newman Lane, Alton, Hants, GU34 2PG.

144   'Electrical safety in arc welding', HS(G) 118, HSE Books, 1994.

145   BS EN 60974: Part 1: 1998 IEC 60974: Part 1: 1998, Arc welding equipment, welding power sources, BSI.

146   BS 638: Part 9: 1990, EN 50060: 1989, Arc welding power sources, equipment and accessories. Specification for power sources for manual arc welding with limited duty, BSI.

147   EW1: 1999 Electric arc welding power sources, NEMA.

148   BS 638: Part 4: 1996 Arc welding power sources, equipment and accessories. Specification for welding cables, BSI.

149   BS EN 60974: Part 12: 1996 Arc welding equipment, coupling devices for welding cables, BSI.

150   BS 638: Part 5: 1988 Arc welding power sources, equipment and accessories. Specification for accessories. BSI.

151 BS EN 638: Part 7: 1984 Arc welding power sources, equipment and accessories. Specification for safety requirements for installation and use, BSI.

152 BS EN 60974: Part 11: 1996 Arc welding equipment. Electrode holders. BSI.

153 'Legionnaire's disease. The control of legionella bacteria in water systems', L8, HSE Books, 2000.

154 BS 499: Part 1: 1991 Welding terms and symbols. Glossary for welding, brazing and thermal cutting, BSI.

155 C1.1: 2000 Recommended practices for resistance welding, AWS.

156 NFPA 70, National electrical code, NFPA, 1999.

157 BS 5924 1989, EN 50063: 1989 Specification for safety requirements for the construction and installation of equipment for resistance welding and allied processes, BSI.

158 C7.1 Recommended practices for electron beam welding, AWS, 1999.

159 N43.3 General safety standards for installations using non-medical X-ray and sealed gamma ray sources of energies up to 10 MeV, ANSI.

160 C7.3 Process specification for electron beam welding, AWS.

161 BS EN 954: Part 1: 1997 Safety of machines. Safety related parts of control systems. General principles of design. BSI.

162 General requirements for all machines, 29 CFR 1910.212.

163 PD 5304: 2000 Safe use of machinery, BSI.

164 Concepts and techniques of machine safeguarding, OSHA document 3067, 1992.

165 Z136.1 American National Standard for the safe use of lasers, Laser Institute of America, ANSI, 2000.

166 BS EN 60825: Part 1: Safety of laser products part 1. Equipment classification, requirements and user's guide, BSI, 1994.

167 BS EN 60825: Part 2, Safety of laser products part 2. Safety of optical fibre communication systems, BSI, 2000.

168 BS EN 207: 1994 Filters and equipment used for personal eye protection against laser radiation, BSI.

169 BS EN 208: 1994 Personal eye protectors used for adjustment work on lasers and laser systems, BSI.

170 Cadmium, 29 CFR 1910.1027.

171 Code of practice for the safe operation of thermal spray equipment, TSSEA, 2001.

172 BS EN 60079: Part 10: Electrical apparatus for explosive gas atmospheres: Classification of hazardous areas, BSI, 1996.

173 'The cleaning and gas freeing of tanks containing flammable residues', CS 15, HSE Books, 1985.

174 F4.1: 1999 Recommended safe practices for the preparation for welding and cutting of containers and piping, AWS.

175 NFPA 326, Standard for the safeguarding of tanks and containers for entry, cleaning or repair, NFPA, 1999.

176 'Safe work in confined spaces. Confined spaces regulations 1997. Approved code of practice, regulations and guidance', L101, HSE Books.

177 Permit required confined spaces, 29 CFR 1910.146.

178 NFPA 53 Fire hazards in oxygen enriched atmospheres, NFPA, 1999.

179 SB2 Oxygen deficient atmospheres, CGA, 2001.

180 BS EN 60529: 1992 Specification for degree of protection provided by enclosures, BSI.

181 Management of Health and Safety at Work Regulations 1999, SI no. 3242, HMSO.

182 Health and Safety (Safety Signs and Signals) Regulations, 1996, SI no. 341, HMSO.

183 Pressure Equipment Regulations 1999, SI 2001, HMSO.

184 Pressure Systems Safety Regulations 2000, SI 128, HMSO.

185 Highly Flammable Liquids and Liquefied Petroleum Gases Regulations 1972, SI 917, HMSO.

186 http://www.osha-slc.gov/OshStd_toc/OSHA_Std_toc_1910.html. The standards can be printed out free of charge. Alternatively they can be purchased from the US Government Printing Office, and enquiries can be made from the Office of Congressional Legislative and Public Affairs, Government Printing Office, 732 North Capitol St NW, Washington, DC 20401.

Index